Recent Advances in Robust Statistics: Theory and Applications

Claudio Agostinelli · Ayanendranath Basu
Peter Filzmoser · Diganta Mukherjee
Editors

Recent Advances in Robust Statistics: Theory and Applications

Springer

Editors
Claudio Agostinelli
Department of Mathematics
University of Trento
Trento, Italy

Peter Filzmoser
Institute of Statistics and Mathematical Methods in Economics
Vienna University of Technology
Vienna, Austria

Ayanendranath Basu
Interdisciplinary Statistical Research Unit
Indian Statistical Institute
Kolkata, India

Diganta Mukherjee
Sampling and Official Statistics Unit
Indian Statistical Institute
Kolkata, India

ISBN 978-81-322-3641-2 ISBN 978-81-322-3643-6 (eBook)
DOI 10.1007/978-81-322-3643-6

Library of Congress Control Number: 2016951695

Printed on acid-free paper

This Springer imprint is published by Springer Nature
The registered company is Springer (India) Pvt. Ltd.
The registered company address is: 7th Floor, Vijaya Building, 17 Barakhamba Road, New Delhi 110 001, India

Preface

This proceedings volume entitled "Recent Advances in Robust Statistics: Theory and Applications" outlines the ongoing research in some topics of robust statistics. It can be considered as an outcome of the International Conference on Robust Statistics (ICORS) 2015, which was held during January 12–16, 2015, at the Indian Statistical Institute in Kolkata, India. ICORS 2015 was the 15th conference in this series, which intends to bring together researchers and practitioners interested in robust statistics, data analysis and related areas. The ICORS meetings create a forum to discuss recent progress and emerging ideas in statistics and encourage informal contacts and discussions among all the participants. They also play an important role in maintaining a cohesive group of international researchers interested in robust statistics and related topics, whose interactions transcend the meetings and endure year round. Previously the ICORS meetings were held at the following places: Vorau, Austria (2001); Vancouver, Canada (2002); Antwerp, Belgium (2003); Beijing, China (2004); Jyväskylä, Finland (2005); Lisbon, Portugal (2006); Buenos Aires, Argentina (2007); Antalya, Turkey (2008); Parma, Italy (2009); Prague, Czech Republic (2010); Valladolid, Spain (2011); Burlington, USA (2012); St. Petersburg, Russia (2013); and Halle, Germany (2014).

More than 100 participants attended ICORS 2015. The scientific program included 80 oral presentations. This program had been prepared by the scientific committee composed of Claudio Agostinelli (Italy), Ayanendranath Basu (India), Andreas Christmann (Germany), Luisa Fernholz (USA), Peter Filzmoser (Austria), Ricardo Maronna (Argentina), Diganta Mukherjee (India), and Elvezio Ronchetti (Switzerland). Aspects of Robust Statistics were covered in the following areas: robust estimation for high-dimensional data, robust methods for complex data, robustness based on data depth, robust mixture regression, robustness in functional data and nonparametrics, statistical inference based on divergence measures, robust dimension reduction, robust methods in statistical computing, non-standard models in environmental studies and other miscellaneous topics in robustness.

Taking advantage of the presence of a large number of experts in robust statistics at the conference, the authorities of the Indian Statistical Institute, Kolkata, and the conference organizers arranged a one-day pre-conference tutorial on robust statistics for the students of the institute and other student members of the local statistics community. Professor Elvezio Ronchetti, Prof. Peter Filzmoser, and Dr. Valentin Todorov gave the lectures at this tutorial class. All the attendees highly praised this effort.

All the papers submitted to these proceedings have been anonymously refereed. We would like to express our sincere gratitude to all the referees. A complete list of referees is given at the end of the book.

This book contains ten articles which we have organized alphabetically according to the first author's name. The paper of Adelchi Azzalini, keynote speaker at the conference, discusses recent developments in distribution theory as an approach to robustness. M. Baragilly and B. Chakraborty dedicate their work to identifying the number of clusters in a data set, and they propose to use multivariate ranks for this purpose. C. Croux and V. Öllerer use rank correlation measures, like Spearman's rank correlation, for robust and sparse estimation of the inverse covariance matrix. Their approach is particularly useful for high-dimensional data. The paper of F.Z. Doğru and O. Arslan examines the mixture regression model, where robustness is achieved by mixtures of different types of distributions. A.-L. Kißlinger and W. Stummer propose scaled Bregman distances for the design of new outlier- and inlier-robust statistical inference tools. A.K. Laha and Pravida Raja A.C. examine the standardized bias robustness properties of estimators when the underlying family of distributions has bounded support or bounded parameter space with applications in circular data analysis and control charts. Large data with high dimensionality are addressed in the contribution of E. Liski, K. Nordhausen, H. Oja, and A. Ruiz-Gazen. They use weighted distances between subspaces resulting from linear dimension reduction methods for combining subspaces of different dimensions. In their paper, J. Miettinen, K. Nordhausen, S. Taskinen, and D.E. Tyler focus on computational aspects of symmetrized M-estimators of scatter, which are multivariate M-estimators of scatter computed on the pairwise differences of the data. A robust multilevel functional data method is proposed by H.L. Shang and applied in the context of mortality and life expectancy forecasting. Highly robust and efficient tests are treated in the contribution of G. Shevlyakov, and the test stability is introduced as a new indicator of robustness of tests.

We would like to thank all the authors for their work, as well as all referees for sending their reviews in time.

Trento, Italy — Claudio Agostinelli
Kolkata, India — Ayanendranath Basu
Vienna, Austria — Peter Filzmoser
Kolkata, India — Diganta Mukherjee
April 2016

Contents

About the Editors

Claudio Agostinelli is Associate Professor of Statistics at the Department of Mathematics, University of Trento, Italy. He received his Ph.D. in Statistics from the University of Padova, Italy, in 1998. Prior to joining the University of Trento, he was Associate Professor at the Department of Environmental Sciences, Informatics and Statistics, Ca' Foscari University of Venice, Italy. His principal area of research is robust statistics. He also works on statistical data depth, circular statistics and computational statistics with applications to paleoclimatology and environmental sciences. He has published over 35 research articles in international refereed journals. He is associate editor of *Computational Statistics*. He is member of the ICORS steering committee.

Ayanendranath Basu is Professor at the Interdisciplinary Statistical Research Unit of the Indian Statistical Institute, Kolkata, India. He received his M.Stat. from the Indian Statistical Institute, Kolkata, in 1986, and his Ph.D. in Statistics from the Pennsylvania State University in 1991. Prior to joining the Indian Statistical Institute, Kolkata, he was Assistant Professor at the Department of Mathematics, University of Texas at Austin, USA. Apart from his primary interest in robust minimum distance inference, his research areas include applied multivariate analysis, categorical data analysis, statistical computing, and biostatistics. He has published over 90 research articles in international refereed journals and has authored and edited several books and book chapters. He is a recipient of the C.R. Rao National Award in Statistics given by the Government of India. He is a Fellow of the National Academy of Sciences, India, and the West Bengal Academy of Science and Technology. He is a past editor of *Sankhya*, The Indian Journal of Statistics, Series B.

Peter Filzmoser studied applied mathematics at the Vienna University of Technology, Austria, where he also wrote his doctoral thesis and habilitation. His research led him to the area of robust statistics, resulting in many international collaborations and various scientific papers in this area. He has been involved in organizing several scientific events devoted to robust statistics, including the first ICORS conference in 2001 in Austria. Since 2001, he has been Professor at the

Department of Statistics at the Vienna University of Technology, Austria. He was Visiting Professor at the Universities of Vienna, Toulouse, and Minsk. He has published over 100 research articles, authored five books and edited several proceedings volumes and special issues of scientific journals. He is an elected member of the International Statistical Institute.

Diganta Mukherjee holds M.Stat. and then Ph.D. (Economics) degrees from the Indian Statistical Institute, Kolkata. His research interests include welfare and development economics and finance. Previously he was a faculty in the Jawaharlal Nehru University, India, Essex University, UK, and the ICFAI Business School, India. He is now a faculty at the Indian Statistical Institute, Kolkata. He has over 60 publications in national and international journals and has authored three books. He has been involved in projects with large corporate houses and various ministries of the Government of India and the West Bengal government. He is acting as a technical advisor to MCX, RBI, SEBI, NSSO, and NAD (CSO).

Flexible Distributions as an Approach to Robustness: The Skew-*t* Case

Adelchi Azzalini

1 Flexible Distributions and Adaptive Tails

1.1 Some Early Proposals

The study of parametric families of distributions with high degree of flexibility, suitable to fit a wide range of shapes of empirical distributions, has a long-standing tradition in statistics; for brevity, we shall refer to this context with the phrase 'flexible distributions'. An archetypal exemplification is provided by the Pearson system with its 12 types of distributions, but many others could be mentioned.

Recall that, for non-transition families of the Pearson system as well as in various other formulations, a specific distribution is identified by four parameters. This allows us to regulate separately from each other four qualitative aspects of a distribution, namely location, scale, slant and tail weight. In the context of robust methods, the appealing aspect of flexibility is represented by the possibility of regulating the tail weight of a continuous distribution to accommodate outlying observations.

When a continuous variable of interest spans the whole real line, an interesting distribution is the one with density function

$$c_\nu \exp\left(-\frac{|x|^\nu}{\nu}\right), \qquad x \in \mathbb{R}, \tag{1}$$

where $\nu > 0$ and the normalizing constant is $c_\nu = \{2\,\nu^{1/\nu}\,\Gamma(1+1/\nu)\}^{-1}$. Here the parameter ν manoeuvres the tail weight in the sense that $\nu = 2$ corresponds to the normal distribution, $0 < \nu < 2$ produces tails heavier than the normal ones, $\nu > 2$ produces lighter tails. The original expression of the density put forward by Subbotin (1923) was set in a different parameterization, but this does not affect our discussion.

A. Azzalini (✉)
Department of Statistical Sciences, University of Padua, Padua, Italy
e-mail: adelchi.azzalini@unipd.it

C. Agostinelli et al. (eds.), *Recent Advances in Robust Statistics: Theory and Applications*, DOI 10.1007/978-81-322-3643-6_1

This flexibility of tail weight provides the motivation for Box and Tiao (1962), Box and Tiao (1973, Sect. 3.2.1), within a Bayesian framework, to adopt the Subbotin's family of distributions, complemented with a location parameter μ and a scale parameter σ, as the parametric reference family allows for departure from normality in the tail behaviour. This logic provides a form of robustness in inference on the parameters of interest, namely μ and σ, since the tail weight parameter adjusts itself to non-normality of the data. Strictly speaking, they consider only a subset of the whole family (1), since the role of ν is played by the non-normality parameter $\beta \in (-1, 1]$ whose range corresponds to $\nu \in [1, \infty)$ and $\beta = 0$ corresponds to $\nu = 2$.

Another formulation with a similar, and even more explicit, logic is the one of Lange et al. (1989). They work in a multivariate context and the error probability distribution is taken to be the Student's t distribution, where the tail weight parameter ν is constituted by the degrees of freedom. Again the basic distribution is complemented by a location and a scale parameter, which are now represented by a vector μ and a symmetric positive-definite matrix, possibly parametrized by some lower dimensional parameter, say ω. Robustness of maximum likelihood estimates (MLEs) of the parameters of interest, μ and ω, occurs "in the sense that outlying cases with large Mahalanobis distances [...] are downweighted", as visible from consideration of the likelihood equations.

The Student's t family allows departures from normality in the form of heavier tails, but does not allow lighter tails. However, in a robustness context, this is commonly perceived as a minor limitation, while there is the important advantage of closure of the family of distributions with respect to marginalization, a property which does not hold for the multivariate version of Subbotin's distribution (Kano 1994).

The present paper proceeds in a similar conceptual framework, with two main aims: (a) to include into consideration also more recent and general proposals of parametric families, (b) to discuss advantages and disadvantages of this approach compared to canonical methods of robustness. For simplicity of presentation, we shall confine our discussion almost entirely to the univariate context, but the same logic carries on in the multivariate case.

1.2 Flexibility via Perturbation of Symmetry

In more recent years, much work has been devoted to the construction of highly flexible families of distributions generated by applying a perturbation factor to a 'base' symmetric density. More specifically, in the univariate case, a density f_0 symmetric about 0 can be modulated to generate a new density

$$f(x) = 2\, f_0(x)\, G_0\{w(x)\}, \qquad x \in \mathbb{R}, \tag{2}$$

for any odd function $w(x)$ and any continuous distribution function G_0 having density symmetric about 0. By varying the ingredients w and G_0, a base density f_0 can give rise to a multitude of new densities f, typically asymmetric but also of more varied shapes. A recent comprehensive account of this formulation, inclusive of its multivariate version, is provided by Azzalini and Capitanio (2014).

One use of mechanism (2) is to introduce asymmetric versions of the Subbotin and Student's t distributions via the modulation factor $G_0\{w(x)\}$. Consider specifically the case when the base density is taken to be the Student's t on ν degrees of freedom, that is,

$$t(x;\nu)=\frac{\Gamma((\nu+1)/2)}{\sqrt{\pi\,\nu}\,\Gamma(\nu/2)}\left(1+\frac{x^2}{\nu}\right)^{-(\nu+1)/2}, \qquad x\in\mathbb{R}. \tag{3}$$

In principle, the choice of the factor $G_0\{w(x)\}$ is bewildering wide, but there are reasons for focusing on the density, denoted as skew-t (ST for short),

$$t(x;\alpha,\nu)=2\,t(x;\nu)\,T\left(\alpha\,x\sqrt{\frac{\nu+1}{\nu+x^2}};\,\nu+1\right), \tag{4}$$

where $T(\cdot;\rho)$ represents the distribution function of a t variate with ρ degrees of freedom and $\alpha\in\mathbb{R}$ is a parameter which regulates slant; $\alpha=0$ gives back the original Student's t. Density (4) is displayed in Fig. 1 for a few values of ν and α.

We indicate only one of the reasons leading to the apparently peculiar final factor of (4). Start by a continuous random variable Z_0 of skew-normal type, that is, with density function

$$\varphi(x;\alpha)=2\,\varphi(x)\,\Phi(\alpha\,x), \qquad x\in\mathbb{R} \tag{5}$$

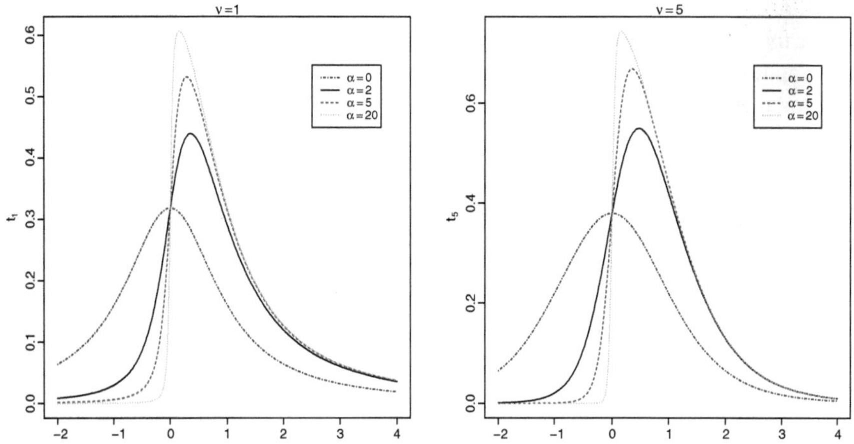

Fig. 1 Skew-t densities when $\nu=1$ in the left plot and $\nu=5$ in the right plot. For each plot, various values of α are considered with $\alpha\geq 0$; the corresponding negative values of α mirror the curves on the opposite side of the vertical axis

where φ and Φ denote the N(0, 1) density and distribution function. An overview of this distribution is provided in Chap. 2 of Azzalini and Capitanio (2014). Consider further $V \sim \chi^2_\nu/\nu$, independent of Z_0, and the transformation $Z = Z_0/\sqrt{V}$, traditionally applied with $Z_0 \sim \mathrm{N}(0, 1)$ to obtain the classical t distribution (3). On assuming instead that Z_0 is of type (5), it can be shown that Z has distribution (4).

For practical work, we introduce location and scale parameters via the transformation $Y = \xi + \omega Z$, leading to a distribution with parameters $(\xi, \omega, \alpha, \nu)$; in this case we write

$$Y \sim \mathrm{ST}(\xi, \omega^2, \alpha, \nu)\,. \tag{6}$$

Because of asymmetry of Z, here ξ does not coincide with the mean value μ; similarly, ω does not equal the standard deviation σ. Actually, a certain moment exists only if ν exceeds the order of that moment, like for an ordinary t distribution. Provided $\nu > 4$, there are known expressions connecting $(\xi, \omega, \alpha, \nu)$ with $(\mu, \sigma, \gamma_1, \gamma_2)$, where the last two elements denote the third and fourth standardized cumulants, commonly taken to be the measures of skewness and excess kurtosis. Inspection of these measures indicates a wide flexibility of the distribution as the parameters vary; notice however that the distribution can be employed also with $\nu \leq 4$, and actually low values of ν represent an interesting situation for applications. Mathematical details omitted here and additional information on the ST distribution are provided in Sects. 4.3 and 4.4 of Azzalini and Capitanio (2014).

Clearly, expression (2) can also be employed with other base distributions and another such option is distribution (1), as expounded in Sect. 4.2 of Azzalini and Capitanio (2014). We do not dwell in this direction because (i) conceptually the underlying logical frame is the same of the ST distribution and (ii) there is a mild preference for the ST proposal. One of the reasons for this preference is similar to the one indicated near the end of Sect. 1.1 in favour of the symmetric t distribution, which is closed under marginalization in the multivariate case and this fact carries on for the ST distribution. Azzalini and Genton (2008) and Sect. 4.3.2 of Azzalini and Capitanio (2014) provide a more extensive discussion of this issue, including additional arguments.

To avoid confusion, the reader must be aware of the existence of other distributions named skew-t in the literature. The one considered here was, presumably, the first construction with this name. The original expression of the density by Branco and Dey (2001) appeared different, since it was stated in an integral form, but subsequently proved by Azzalini and Capitanio (2003) to be equivalent to (3).

The high flexibility of these distributions, specifically the possibility to regulate their tail weight combined with asymmetry, supports their use in the same logic of the papers recalled in Sect. 1.1. Azzalini (1986) has motivated the introduction of asymmetric versions of Subbotin distribution precisely by robustness considerations, although this idea has not been complemented by numerical exploration. Azzalini and Genton (2008) have worked in a similar logic, but focusing mainly on the ST distribution as the working reference distribution; more details are given in Sect. 3.4.

To give a first perception of the sort of outcome to be expected, let us consider a very classical benchmark of robustness methodology, perhaps *the* most classical:

Table 1 Total absolute deviation of various fitting methods applied to the stack loss data

Method	LS	Huber	LTS	MM	MLE-ST
Q	49.7	46.1	49.4	45.3	43.4

the 'stack loss' data. We use the data following the same scheme of many existing publications, by fitting a linear regression model with the three available explanatory variables plus intercept to the response variable y, i.e. the stack loss, and examine the discrepancy between observed and fitted values along the $n = 21$ data points. A simple measure of the achieved goodness of fit is represented by the total absolute deviation

$$Q = \sum_{i=1}^{n} |y_i - \hat{y}_i|,$$

where y_i denotes the ith observation of the response variable and $\hat{y}_i$ is the corresponding fitted value produced by any candidate method. The methods considered are the following: least squares (LS, in short), Huber estimator with scale parameter estimated by minimum absolute deviation, least trimmed sum of squares (LTS) of Rousseeuw and Leroy (1987), MM estimation proposed by Yohai (1987), MLE under assumption of ST distribution of the error term (MLE-ST). For the ST case, an adjustment to the intercept must be made to account for the asymmetry of the distribution; here we have added the median of the fitted ST error distribution to the crude estimate of the intercept. The outcome is reported in Table 1, whose entries have appeared in Table 5 of Azzalini and Genton (2008) except that MM estimation was not considered there. The Q value of MLE-ST is the smallest.

2 Aspects of Robustness

2.1 *Robustness and Real Data*

The effectiveness of classical robust methods in work with real data has been questioned in a well-known paper by Stigler (1977). In the opening section, the author lamented that 'most simulation studies of the robustness of statistical procedures have concentrated on a rather narrow range of alternatives to normality: independent, identically distributed samples from long-tailed symmetric continuous distributions' and proposed instead 'why not evaluate the performance of statistical procedures with *real* data?' He then examined 24 data sets arising from classical experiments, all targeted to measure some physical or astronomical quantity, for which the modern measurement can be regarded as the true value. After studying these data sets, including application of a battery of 11 estimators on each of them, the author concluded in the final section that 'the data sets examined do exhibit a slight tendency towards more extreme values that one would expect from normal samples, but a very small amount of trimming seems to be the best way to deal with this. [...] The more drastic modern remedies for feared gross errors [...] lead here to an unnecessary loss of efficiently.'

Similarly, Hill and Dixon (1982) start by remarking that in the robustness literature 'most estimators have been developed and evaluated for mathematically well-behaved symmetric distributions with varying degrees of high tail', while 'limited consideration has been given to asymmetric distributions'. Also in this paper the programme is to examine the distribution of really observed data, in this case originating in an clinical laboratory context, and to evaluate the behaviour of proposed methods on them. Specifically, the data represent four biomedical variables recorded on '3000 apparently well visitors' of which, to obtain a fairly homogeneous population, only data from women 20–50 years old were used, leading to sample sizes in the range 1037–1110 for the four variables. Also for these data, the observed distributions 'differ from many of the generated situations currently in vogue: the tails of the biomedical distributions are not so extreme, and the densities are often asymmetric, lumpy and have relatively few unique values'. Other interesting aspects arise by repeatedly extracting subsamples of size 10, 20 and 40 from the full set, computing various estimators on these subsamples and examining the distributions of the estimators. The indications that emerge include the fact that the population values of the robust estimators do not estimate the population mean; moreover, as the distributions become more asymmetric, the robust estimates approach the population median, moving away from the mean.

A common indication from the two above-quoted papers is that the observed distributions display some departure from normality, but tail heaviness is not as extreme as in many simulation studies of the robustness literature. The data display instead other forms of departures from ideal conditions for classical methods, especially asymmetry and "lumpiness" or granularity. However, the problem of granularity will be presumably of decreasing importance as technology evolves, since data collection takes place more and more frequently in an automated manner, without involving manual transcription and consequent tendency to number rounding, as it was commonly the case in the past.

Clearly, these indications must not be regarded as universal. Stigler (1977, Sect. 6) himself recognizes that 'some real data sets with symmetric heavy tails do exist, cannot be denied'. In addition, it can be remarked that the data considered in the quoted papers are all of experimental or laboratory origin, and possibly in a social sciences context the picture may be somewhat different. However, at the least, the indication remains that the distribution of real data sets is not systematically symmetric and not so heavy tailed as one could perceive from the simulation studies employed in a number of publications.

2.2 Some Qualitative Considerations

The plan of this section is to discuss qualitatively the advantages and limitation of the proposed approach, also in the light of the facts recalled in the preceding subsection.

For the sake of completeness, let us state again and even more explicitly the proposed line of work. For the estimation of parameters of interest in a given inferential problem, typically location and scale, we embed them in a parametric class which includes some additional parameters capable of regulating the shape and tail behaviour of the distribution, so to accommodate outlying observations as manifestations of the departures from normality of these distributions, hence providing a form of robustness. In a regression context, the location parameter is replaced by the regression parameters as the focus of primary interest.

In this logic, an especially interesting family of distributions is the skew-t, which allows to regulate both its asymmetry and tail weight, besides location and scale. Such a usage of the distribution was not the original motivation of its design, which was targeted to flexibility to adapt itself to a variety of situations, but this flexibility leads naturally to this other role.

The formulation prompts a number of remarks, in different and even contrasting directions, partly drawing from Azzalini and Genton (2008) and from Azzalini and Capitanio (2014, Sect. 4.3.5).

1. Clearly the proposed route does not belong to the canonical formulation of robust methods, as presented for instance by Huber and Ronchetti (2009), and one cannot expect it to fulfil the criteria stemming from that theory. However, some connections exist. Hill and Dixon (1982, Sect. 3.1) have noted that the Laszlo robust estimator of location coincides with the MLE for the location parameter of a Student's t when its degrees of freedom are fixed. Lucas (1997), He et al. (2000) examine this connection in more detail, confirming the good robustness properties of MLE of the location parameter derived from an assumption of t distribution with fixed degrees of freedom.
2. The key motivation for adopting the flexible distributions approach is to work with a fully specified parametric model. Among the implied advantages, an important one is that it is logically clear what the estimands are: the parameters of the model. The same question is less transparent with classical robust methods. For the important family of M-estimators, the estimands are given implicitly as the solution of a certain nonlinear equation; see for instance Theorem 6.4 of Huber and Ronchetti (2009). In the simple case of a location parameter estimated using an odd ψ-function when the underlying distribution is symmetric around a certain value, the estimand is that centre of symmetry, but in a more general setting we are unable to make a similarly explicit statement.
3. Another advantage of a fully specified parametric model is that, at the end of the inference process, we obtain precisely that, a fitted probability model. Hence, as a simple example, one can assess the probability that a variable of interest lies in a given interval (a, b), a question which cannot be tackled if one works with estimating equations as with M-estimates.

4. The critical point for a parametric model is of course the inclusion of the true distribution underlying the data generation among those contemplated by the model. Since models can only approximate reality, this ideal situation cannot be met exactly in practice, except exceptional situations. If we denote by $\theta \in \Theta \subseteq \mathbb{R}^p$ the parameter of a certain family of distributions, $f(x;\theta)$, recall that, under suitable regularity conditions, the MLE $\hat{\theta}$ of θ converges in probability to the value $\theta_0 \in \Theta$ such that $f(x;\theta_0)$ has minimal Kullback–Leibler divergence from the true distribution. The approach via flexible distributions can work satisfactorily insofar it manages to keep this divergence limited in a wide range of cases.
5. Classical robust methods are instead designed to work under all possible situations, even the most extreme. On the other hand, empirical evidence recalled in Sect. 2.1 indicates that protection against all possible alternatives may be more than we need, as in the real world the most extreme situations do not arise that often.
6. As for the issue discussed in item 4, we are not disarmed, because the adequacy of a parametric model can be tested a posteriori using model diagnostic tools, hence providing a safeguard against appreciable Kullback–Leibler divergence.

3 Some Quantitative Indications

The arguments presented in Sect. 2.2, especially in items 4 and 5 of the list there, call for quantitative examination of how the flexible distribution approach works in specific cases, especially when the data generating distributions does not belong to the specified parametric distribution, and how it compares with classical robust methods.

This is the task of the present section, adopting the ST parametric family (6) and using MLE for estimation; for brevity we refer to this option as MLE-ST. Notice that ν is not fixed in advance, but estimated along with the other parameters. When a similar scheme is adopted for the classical Student's t distribution, Lucas (1997) has shown that the influence function becomes unbounded, hence violating the canonical criteria for robustness. A similar fact can be shown to happen with the ST distribution.

3.1 Limit Behaviour Under a Mixture Distribution

Recall the general result about the limit behaviour of the MLE when a certain parametric assumption is made on the distribution of an observed random variable Y, whose actual distribution $p(\cdot)$ may not be a member of the parametric class. Under the assumption of independent sampling from Y with constant distribution p and various regularity conditions, Theorem 2 of Huber (1967) states that the MLE of parameter θ converges almost surely to the solution θ_0, assumed to be unique, of the equation

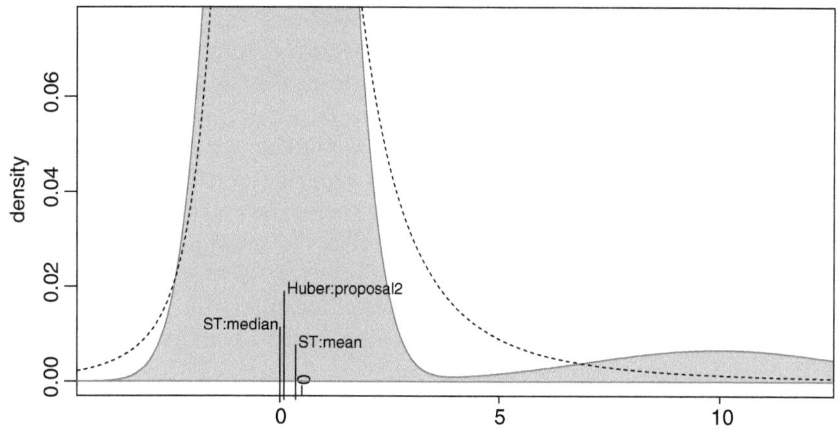

Fig. 2 The shaded area represents the main body of distribution (8) when $\pi = 0.05$, $\Delta = 10$, $\sigma = 3$ and the *small circle* on the horizontal axis marks its mean value; the *dashed curve* represents the corresponding MLE-ST limit distribution. The *vertical bars* denote the estimands of Huber's 'proposal 2' and of MLE-ST, the latter one in two variants, mean value and median

$$\mathbb{E}_p\{\psi(Y;\theta)\} = 0, \tag{7}$$

where the subscript p indicates that the expectation is taken with respect to that distribution and $\psi(\cdot;\theta)$ denotes the score function of the parametric model.

We examine numerically the case where the parametric assumption is of ST type with $\theta = (\xi, \omega, \alpha, \nu)$ and $p(x)$ represents a contaminated normal distribution, that is, a mixture density of the form

$$p(x) = (1-\pi)\,\varphi(x) + \pi\,\sigma^{-1}\,\varphi\{\sigma^{-1}(x-\Delta)\}\,. \tag{8}$$

In our numerical work, we have set $\pi = 0.05$, $\Delta = 10$, $\sigma = 3$. The corresponding $p(x)$ is depicted as a grey-shaded area in Fig. 2 and its mean value, 0.5, is marked by a small circle on the horizontal axis. The expression of the four-dimensional score function for the ST assumption is given by DiCiccio and Monti (2011), reproduced with inessential changes of notation in Sect. 4.3.3 of Azzalini and Capitanio (2014). The solution of (7) obtained via numerical methods is $\theta_0 = (-0.647, 1.023, 1.073, 2.138)$, whose corresponding ST density is represented by the dashed curve in Fig. 2. From θ_0, we can compute standard measures of location, such as the mean and the median of the ST distribution with that parameter; their values, 0.0031 and 0.3547, are marked by vertical bars on the plot. The first of these values is almost equal to the centre of the main component of $p(x)$, i.e. $\varphi(x)$, while the mean of the ST distribution is not far from the mean of $p(x)$. Which of the two quantities is more appropriate to consider depends, at least partly, on the specific application under consideration.

To obtain a comparison term from a classical robust technique, a similar numerical evaluation has been carried out for 'proposal 2' of Huber (1964), where θ comprises a location and a scale parameter. The corresponding estimands are computed solving an equation formally identical to (7), except that now ψ represents the set of estimating equations, not the score function; see Theorem 6.4 of Huber and Ronchetti (2009). For the case under consideration, the location estimand is 0.0957, which is also marked by a vertical bar in Fig. 2. This value is intermediate to the earlier values of the ST distribution, somewhat closer to the median, but anyway they are all not far away from each other.

For the ST distribution, alternative measures of location, scale and so on, which are formally similar to the corresponding moment-based quantities but exist for all $\nu > 0$, have been proposed by Arellano-Valle and Azzalini (2013). In the present case, the location measure of this type, denoted pseudomean, is equal to 0.1633 which is about halfway the ST mean and median; this value is not marked on Fig. 2 to avoid cluttering.

3.2 A Non-random Simulation

We examine the behaviour of ST-MLE and other estimators when an "ideal sample" is perturbed by suitably modifying one of its components. As an ideal sample we take the vector $z_1, \ldots, z_n$, where z_i denotes the expected value of the ith order statistics of a random sample of size n drawn from the N(0, 1) distribution, and its perturbed version has ith component as follows:

$$y_i = \begin{cases} z_i & \text{if } i = 1, \ldots, n-1, \\ z_n + \Delta & \text{if } i = n. \end{cases}$$

For any given $\Delta > 0$, we examine the corresponding estimates of location obtained from various estimation methods and then repeat the process for an increasing sequence of displacements Δ. Since the y_i's are artificial data, the experiment represents a simulation, but no randomness is involved. Another way of looking at this construction is as a variant form of the sensitivity curve.

In the subsequent numerical work, we have set $n = 100$, so that $-2.5 < z_i < 2.5$, and Δ ranges from 0 to 15. Computation of the MLE for the ST distribution has been accomplished using the R package `sn` (Azzalini 2015), while support for classical robust procedures is provided by packages `robust` (Wang et al. 2014) and `robustbase` (Rousseeuw et al. 2014); these packages have been used at their default settings. The degrees of freedom of the MLE-ST fitted distributions decrease from about 4×10^4 (which essentially is a numerical substitute of ∞) when $\Delta = 0$, down to $\hat{\nu} = 3.57$ when $\Delta = 15$.

For each MLE-ST fit, the corresponding median, mean value and pseudomean of the distribution have been computed and these are the values plotted in Fig. 3 along with the sample average and some representatives of the classical robust method-

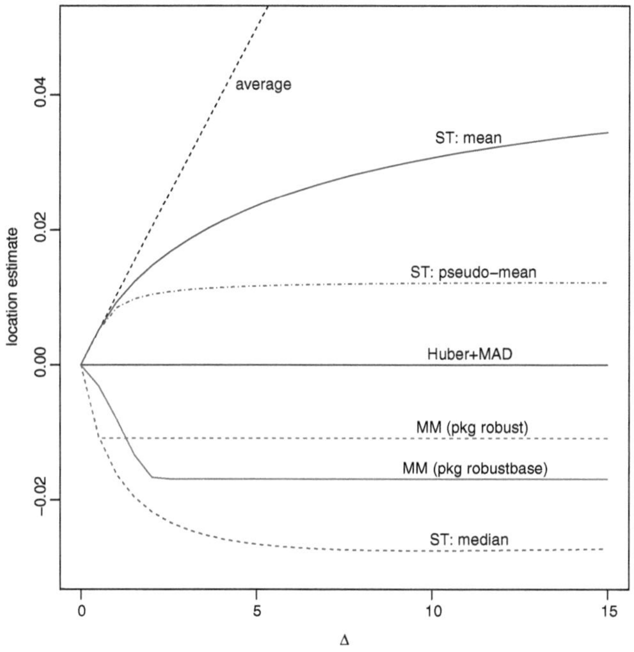

Fig. 3 Estimates of the location parameter applied to a perturbed version of the expected normal order statistics plotted versus the displacement Δ

ology. The slight difference between the two curves of MM estimates is due to a small difference in the tuning parameters of the R packages. Inevitably, the sample average diverges linearly as Δ increases. The ST median and pseudomean behave qualitatively much like the robust methods, while the mean increases steadily, but far more gently than the sample average, following a logarithmic-like sort of curve.

3.3 A Random Simulation

Our last numerical exhibit refers to a regular stochastic simulation. We replicate an experiment where $n = 100$ data points are sampled independently from the regression scheme

$$y = \beta_0 + \beta_1 x + \varepsilon,$$

where the values of x are equally spaced in (0, 10), $\beta_0 = 0$, $\beta_1 = 2$ and the error term ε has contaminated normal distribution of type (8) with $\Delta \in \{2.5, 5, 7.5, 10\}$, $\pi \in \{0.05, 0.10\}$, $\sigma = 3$.

For each generated sample, estimates of β_0 and β_1 have been computed using least squares (LS), least trimmed sum of squared (LTS), MM estimation and MLE-

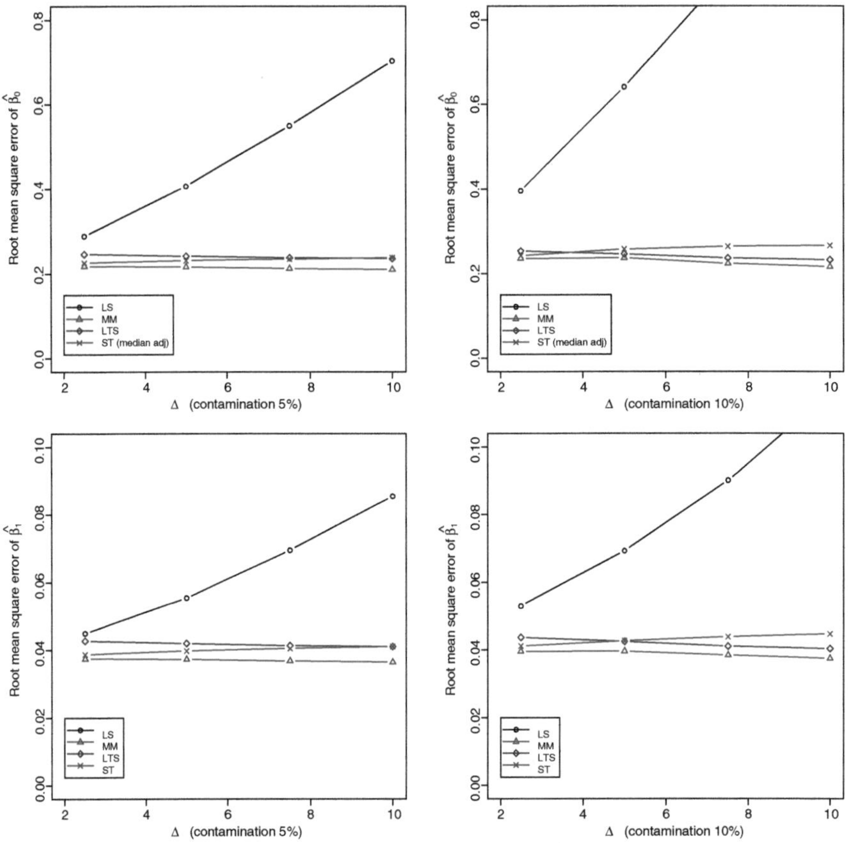

Fig. 4 Root-mean-square error in estimation of β_0 (*top panels*) and β_1 (*bottom*) from a linear regression setting where the error term has contaminated normal distribution with contamination level 5 % (*left*) and 10 % (*right*), as estimated from 50,000 replications [Reproduced with permission from Azzalini and Capitanio (2014)]

ST with median adjustment of the intercept; all of them have already been considered and described in an earlier section. After 50,000 replications of this step, the root-mean-square (RMS) error of the estimates has been computed and the final outcome is presented in Fig. 4 in the form of plots of RMS error versus Δ, separately for each parameter and each contamination level.

The main indication emerging from Fig. 4 is that the MLE-ST procedure behaves very much like the classical robust methods over a wide span of Δ. There is a slight increase of the RMS error of MLE-ST over MM and LTS when we move to the far right of the plots; this is in line with the known non-robustness of MLE-ST with respect to the classical criteria. However, this discrepancy is of modest entity and presumably it would require very large values of Δ to become appreciable. Notice that on the right side of the plots we are already 10 standard deviations away from the centre of $\varphi(x)$, the main component of distribution (8).

3.4 Empirical and Applied Work

The MLE-ST methodology has been tested on a number of real datasets and application areas. A fairly systematic empirical study has been presented by Azzalini and Genton (2008), employing data originated from a range of situations: multiple linear regression, linear regression on time series data, multivariate observations, classification of high dimensional data. Work with multivariate data involves using the multivariate skew-t distribution, of which an account is presented in Chap. 6 of Azzalini and Capitanio (2014). In all the above-mentioned cases, the outcome has been satisfactory, sometimes very satisfactory, and has compared favourably with techniques specifically developed for the different situations under consideration.

Applications of the ST distribution arise in a number of fields. We do not attempt a complete review, but only indicate some directions. One point to bear in mind is that often, in applied work, the distinction between long tails and outlying observations is effectively blurred.

A crystalline exemplification of the last statement is provided by the returns generated in the industry of artistic productions, especially from films and music. Here the so-called 'superstar effect' leads to values of a few isolated units which are far higher than the main body of the production. These extremely large values are outlying but not spurious; they are genuine manifestations of the phenomenon under study, whose probability distribution is strongly asymmetric and heavy tailed, even after log transformation of the original data. See Walls (2005) and Pitt (2010) for a complete discussion and for illustrations of successful use of the ST distribution.

The above-described data pattern and corresponding explorations of use of the MLE-ST procedure exist also in other application areas. Among these, quantitative finance represents a prominent example and this has prompted also significant theoretical contributions to the development of this area; see Adcock (2010, 2014). Another important context is represented by natural phenomena, where occasionally extreme values jump far away from the main body of the observations; applied work in this direction includes multivariate modelling of coastal flooding (Thompson and Shen 2004), monthly precipitations (Marchenko and Genton 2010), riverflow intensity (Ghizzoni et al. 2010, 2012).

Another direction currently under vigorous investigation is model-based cluster analysis. The traditional assumption that each component of the underlying mixture distribution is multivariate normal is often too restrictive, leading to an inappropriate increase of the number of component distributions. A more flexible distribution, such as the multivariate ST, can overcome this limitation, as shown in an early application by Pyne et al. (2009), but various other papers along a similar line exist, including of course adoption of other flexible distributions.

At least a mention is due of methods for longitudinal data and mixed effect models, such as in Lachos et al. (2010), Ho and Lin (2010).

We stress once more that the above-quoted contributions have been picked up as the representatives of a substantially broader collection, which includes additional methodological themes and application areas. A more extensive summary of this activity is provided in the monograph of Azzalini and Capitanio (2014).

In connection with applied work, it is appropriate to underline that care must be exercised in numerical maximization of the likelihood function, at least with certain datasets. It is known that fitting a classical Student's t distribution with unconstrained degrees of freedom can be problematic, especially in the multivariate case; the inclusion of a skewness parameter adds another level of complexity. It is then advisable to start the maximization process from various starting points. In problematic cases, computation of the profile likelihood function with respect to ν can be a useful device. Advancements on the reliability and efficiency of optimization techniques for this formulation would be valuable.

4 Concluding Remarks

The overall message which can be extracted from the preceding pages is that flexible distributions constitute a credible approach to the problem of robustness. Since it does not descend from the canonical scheme of classical robust methods, this approach cannot meet the classical robustness optimality criteria. However, these criteria are targeted to offer protection against extreme situations which in real data are not so commonly encountered, perhaps even seldom encountered. In less extreme situations, but still allowing for appreciable departure from normality, flexible distributions, specially in the representative case of the skew-t distribution, offer adequate protection against problematic situations, while providing a fully specified probability model, with the qualitative advantages discussed in Sect. 2.2.

We have adopted the ST family as our working parametric family, but the reasons for this preference, explained briefly above and more extensively by Azzalini and Genton (2008), are not definitive; in certain problems, it may well be appropriate to work with some other distribution. For instance, if one envisages that the problem under consideration contemplates departure from normality in the form of shorter tails or possibly a combination of longer and shorter tails in different subcases, and the setting is univariate, then the Subbotin distribution and its asymmetric variants represent an interesting option.

Acknowledgments This paper stems directly from my oral presentation with the same title delivered at the ICORS 2015 conference held in Kolkata, India. I am grateful to the conference organizers for the kind invitation to present my work in that occasion. Thanks are also due to attendees at the talk that have contributed to the discussion with useful comments, some of which have been incorporated here.

References

Adcock CJ (2010) Asset pricing and portfolio selection based on the multivariate extended skew-Student-t distribution. Ann Oper Res 176(1):221–234. doi:10.1007/s10479-009-0586-4

Adcock CJ (2014) Mean-variance-skewness efficient surfaces, Stein's lemma and the multivariate extended skew-Student distribution. Eur J Oper Res 234(2):392–401. doi:10.1016/j.ejor.2013.07.011. Accessed 20 July 2013

Arellano-Valle RB, Azzalini A (2013) The centred parameterization and related quantities of the skew-t distribution. J Multiv Anal 113:73–90. doi:10.1016/j.jmva.2011.05.016. Accessed 12 June 2011

Azzalini A (1986) Further results on a class of distributions which includes the normal ones. Statistica XLVI(2):199–208

Azzalini A (2015) The R package sn: The skew-normal and skew-t distributions (version 1.2-1). Università di Padova, Italia. http://azzalini.stat.unipd.it/SN

Azzalini A, Capitanio A (2003) Distributions generated by perturbation of symmetry with emphasis on a multivariate skew t distribution. J R Statis Soc ser B 65(2):367–389, full version of the paper at arXiv.org:0911.2342

Azzalini A with the collaboration of Capitanio A (2014) The Skew-Normal and Related Families. IMS Monographs, Cambridge University Press, Cambridge. http://www.cambridge.org/9781107029279

Azzalini A, Genton MG (2008) Robust likelihood methods based on the skew-t and related distributions. Int Statis Rev 76:106–129. doi:10.1111/j.1751-5823.2007.00016.x

Box GEP, Tiao GC (1962) A further look at robustness via Bayes's theorem. Biometrika 49:419–432

Box GP, Tiao GC (1973) Bayesian inference in statistical analysis. Addison-Wesley Publishing Co

Branco MD, Dey DK (2001) A general class of multivariate skew-elliptical distributions. J Multiv Anal 79(1):99–113

DiCiccio TJ, Monti AC (2011) Inferential aspects of the skew t-distribution. Quaderni di Statistica 13:1–21

Ghizzoni T, Roth G, Rudari R (2012) Multisite flooding hazard assessment in the Upper Mississippi River. J Hydrol 412–413(Hydrology Conference 2010):101–113. doi:10.1016/j.jhydrol.2011.06.004

Ghizzoni T, Roth G, Rudari R (2010) Multivariate skew-t approach to the design of accumulation risk scenarios for the flooding hazard. Adv Water Res 33(10, Sp. Iss. SI):1243–1255. doi:10.1016/j.advwatres.2010.08.003

He X, Simpson DG, Wang GY (2000) Breakdown points of t-type regression estimators. Biometrika 87:675–687

Hill MA, Dixon WJ (1982) Robustness in real life: a study of clinical laboratory data. Biometrics 38:377–396

Ho HJ, Lin TI (2010) Robust linear mixed models using the skew t distribution with application to schizophrenia data. Biometr J 52:449–469. doi:10.1002/bimj.200900184

Huber PJ (1964) Robust estimation of a location parameter. Ann Math Statis 35:73–101. doi:10.1214/aoms/1177703732

Huber PJ (1967) The behaviour of maximum likelihood estimators under nonstandard conditions. In: Le Cam LM, Neyman J (eds) Proceedings of the fifth Berkeley symposium on mathematical statistics and probability, vol 1. University of California Press, pp 221–23

Huber PJ, Ronchetti EM (2009) Robust statistics, 2nd edn. Wiley

Kano Y (1994) Consistency property of elliptical probability density functions. J Multiv Anal 51:139–147

Lachos VH, Ghosh P, Arellano-Valle RB (2010) Likelihood based inference for skew-normal independent linear mixed models. Statist Sinica 20:303–322

Lange KL, Little RJA, Taylor JMG (1989) Robust statistical modeling using the t-distribution. J Am Statis Assoc 84:881–896

Lucas A (1997) Robustness of the Student *t* based M-estimator. Commun Statis Theory Meth 26(5):1165–1182. doi:10.1080/03610929708831974

Marchenko YV, Genton MG (2010) Multivariate log-skew-elliptical distributions with applications to precipitation data. Environmetrics 21(3-4, Sp. Iss.SI):318–340. doi:10.1002/env.1004

Pitt IL (2010) Economic analysis of music copyright: income, media and performances. Springer Science & Business Media. http://www.springer.com/book/9781441963178

Pyne S, Hu X, Wang K, Rossin E, Lin TI, Maier LM, Baecher-Alland C, McLachlan GJ, Tamayo P, Hafler DA, De Jagera PL, Mesirov JP (2009) Automated high-dimensional flow cytometric data analysis. PNAS 106(21):8519–8524. doi:10.1073/pnas.0903028106

Rousseeuw P, Croux C, Todorov V, Ruckstuhl A, Salibian-Barrera M, Verbeke T, Koller M, Maechler M (2014) `robustbase`: basic robust statistics. http://CRAN.R-project.org/package=robustbase, R package version 0.91-1

Rousseeuw PJ, Leroy AM (1987) Robust regression and outlier detection. Wiley, New York

Stigler SM (1977) Do robust estimators work with real data? (with discussion). Ann Statis 5(6):1055–1098

Subbotin MT (1923) On the law of frequency of error. Matematicheskii Sbornik 31:296–301

Thompson KR, Shen Y (2004) Coastal flooding and the multivariate skew-*t* distribution. In: Genton MG (ed) Skew-elliptical distributions and their applications: a journey beyond normality, Chap 14. Chapman & Hall/CRC, pp 243–258

Walls WD (2005) Modeling heavy tails and skewness in film returns. Appl Financ Econ 15(17):1181–1188. doi:10.1080/09603100500391040, http://www.tandf.co.uk/journals

Wang J, Zamar R, Marazzi A, Yohai V, Salibian-Barrera M, Maronna R, Zivot E, Rocke D, Martin D, Maechler M, Konis K (2014) `robust`: robust library. http://CRAN.R-project.org/package=robust, R package version 0.4-16

Yohai VJ (1987) High breakdown-point and high efficiency robust estimates for regression. Ann Statis 15(20):642–656

Determining the Number of Clusters Using Multivariate Ranks

Mohammed Baragilly and Biman Chakraborty

1 Introduction

Most of the clustering algorithms available in the literature require the number of clusters to be fixed a priori. That makes the determination of the number of clusters a very important problem in cluster analysis for multivariate data. Over the last 40 years, a wealth of publications have introduced and discussed many graphical approaches and statistical algorithms to determine the cluster sizes and the number of clusters. However, there is no universally acceptable solution to this problem due to the complexity of the high-dimensional real data sets. It is also well known that using different clustering methods may give different numbers of clusters.

As an example, a biologist would like to find out the clusters from the DNA microarray data on gene expressions, and consequently detecting the classes or sub-classes of diseases. Other common examples are in the research areas of the taxonomy of animals and plants, in construction of phylogenetic trees, handwriting recognition and measuring the similarities of the different languages. A good collection of the cluster analysis examples are available in Hartigan (1975), Gan et al. (2007) and Everitt et al. (2011).

In a model-based clustering approach, one may assume that the d-dimensional data $\mathbf{X}_1, \ldots, \mathbf{X}_n$ are coming from a mixture probability density function

$$f(\mathbf{x}) = p_1 f_1(\mathbf{x}) + \cdots + p_k f_k(\mathbf{x}) \tag{1}$$

M. Baragilly (✉) · B. Chakraborty
School of Mathematics, University of Birmingham, Birmingham B15 2TT, UK
e-mail: MHB110@bham.ac.uk; mohammed.baragilly@gmail.com

B. Chakraborty
e-mail: B.Chakraborty@bham.ac.uk

M. Baragilly
Department of Applied Statistics, Helwan University, Cairo, Egypt

C. Agostinelli et al. (eds.), *Recent Advances in Robust Statistics: Theory and Applications*, DOI 10.1007/978-81-322-3643-6_2

where $f_1, \ldots, f_k$ are d-dimensional unimodal density functions and $p_1, \cdots, p_k$ are the mixing proportions with $p_1 + \cdots + p_k = 1$. In most of the traditional clustering algorithms like k-means, the number of densities k is assumed to be known and the density functions $f_1, \cdots, f_k$ are estimated parametrically. Mixing proportions are estimated differently in different algorithms (Hartigan 1975). However, an important problem is to find the number of clusters k itself. The early works on cluster number determination methods were from Thorndike (1953), Friedman and Rubin (1967), Beale (1969), Marriott (1971), Duda and Hart (1973), Calinski and Harabasz (1974), Hartigan (1975). Along the line of all the previous methods, there are many other attempts and algorithms have been suggested in order to estimate the number of clusters (e.g., Mojena (1977), Krzanowski and Lai (1985), Milligan and Cooper (1985), Kaufman and Rousseeuw (1990), Overall and Magee (1992), Gordon (1998), Tibshirani et al. (2001), Sugar and James (2003)). The focus of these clustering stopping-rules (indices) are on computing some criterion function for each cluster solution and then one chooses the solution that indicates most distinct clustering. Most of these standard approaches depend on within and between cluster variations. In this work, we explore a proposal to determine the number of clusters, k, as well as mixing proportions $p_1, \cdots, p_k$ visually using a forward search algorithm.

The main idea of a forward search algorithm is to grow the cluster size starting from an initial subset of observations based on some kind of distance measure. It plots a statistic against the size of the subset for easy detection of clusters. The traditional forward search approach based on Mahalanobis distances have been introduced by Hadi (1992), Hadi and Simonoff (1993), where they considered a forward search, which terminates when the subset size m is the median of the number of observations, while a similar method used by Atkinson and Mulira (1993), Atkinson (1994) continues until $m = n$, the sample size. Atkinson et al. (2004) introduced many applications of the forward search in the analysis of multivariate data. A good overview of the forward search and its applications is available in Atkinson et al. (2010).

All the previous literature assumed Mahalanobis distance as the distance measure to be used in the forward search procedure. It is well known that Mahalanobis distance is invariant under all nonsingular transformations and it also performs well with the Gaussian mixture models (GMM), however, it cannot be correctly applied to asymmetric distributions and more generally to distributions, which depart from the elliptical symmetry assumptions. In order to address this limitation, in this paper, we propose a new forward search methodology based on spatial ranks and volume of central rank regions (Chaudhuri 1996; Serfling 2002) to tackle the problem of heavy tailed mixture distributions with higher dimensional data. For last two decades, spatial ranks are being used in analyzing multivariate data nonparametrically. They are easy to compute, but do not depend on parameter estimates of the underlying distributions, which make them robust against distributional assumptions. Koltchinskii (1997) also proved that the spatial ranks characterize a multivariate distribution.

The remainder of the paper is organized as follows. In Sect. 2, we introduce the forward search method based on spatial ranks and volume of central rank regions. In Sect. 3.1, we give some numerical examples based on simulated data sets to show the performance of the proposed algorithm when some heavy tailed mixture distributions under the elliptic symmetry case are considered. Section 3.2 demonstrates the results of two real data sets compared to some standard methods. Finally, we present some concluding remarks in Sect. 4.

2 Forward Search with Multivariate Ranks

For $\mathbf{x} \in \mathbb{R}^d$, the multivariate spatial sign function is defined as

$$sign(\mathbf{x}) = \begin{cases} \frac{\mathbf{x}}{\|\mathbf{x}\|} & \text{if } \mathbf{x} \neq 0 \\ 0 & \text{if } \mathbf{x} = 0 \end{cases} \tag{2}$$

where $\|\mathbf{x}\|$ is the Euclidean norm such that; $\|\mathbf{x}\| = \sqrt{x_1^2 + x_2^2 + ... + x_d^2}$. Note that this is nothing but the direction of the d-dimensional vector $\mathbf{x}$.

Suppose that $\mathbf{X} \in \mathbb{R}^d$ has a d-dimensional distribution F, which is assumed to be absolutely continuous throughout this paper, then the multivariate spatial rank function of the point $\mathbf{x} \in \mathbb{R}^d$ with respect to F can be defined as

$$Rank_F(\mathbf{x}) = E_F\left(\frac{\mathbf{x} - \mathbf{X}}{\|\mathbf{x} - \mathbf{X}\|}\right). \tag{3}$$

Now suppose that $\mathbf{X}_1, \mathbf{X}_2, \ldots, \mathbf{X}_n \in \mathbb{R}^d$ is a random sample with distribution F, then the sample version of the multivariate spatial rank function of $\mathbf{x} \in \mathbb{R}^d$ with respect to $\mathbf{X}_1, \mathbf{X}_2, \ldots, \mathbf{X}_n$ is given by

$$Rank_{F_n}(\mathbf{x}) = \frac{1}{n}\sum_{i=1}^{n} Sign(\mathbf{x} - \mathbf{X}_i) = \frac{1}{n}\sum_{i=1}^{n} \frac{\mathbf{x} - \mathbf{X}_i}{\|\mathbf{x} - \mathbf{X}_i\|}. \tag{4}$$

It can be clearly noticed that, when $Rank_F(\mathbf{x}) = \mathbf{0}$, $\mathbf{x}$ is the spatial median and $Rank_F(\mathbf{x}) = \mathbf{u}$ implies that $\mathbf{x}$ is the $\mathbf{u}$th geometric quantile (Chaudhuri 1996) of F.

In the forward search algorithm, let $S(m)$ be a subset of size m at a particular stage. Then define the spatial ranks of individual observations corresponding to the subset $S(m)$ as

$$r_i(m) = \frac{1}{m}\sum_{j \in S(m)} \frac{\mathbf{X}_i - \mathbf{X}_j}{\|\mathbf{X}_i - \mathbf{X}_j\|} \tag{5}$$

for $i = 1, \ldots, n$. Let us now introduce the forward search procedure based on the multivariate spatial ranks.

Forward search algorithm with spatial ranks

1. In order to start the search, we need to choose an initial subset. Suppose that $S(m)$ is the initial subset with $m = d + 1$, then one search can be run from this starting point.
2. Calculate the spatial ranks $r_i(m)$ depending on the observations in the subset $S(m)$.
3. Compute $r_{min}(m)$, where $r_{min}(m) = \min \|r_i(m)\|; i \notin S(m)$.
4. Grow the subset $S(m)$ to $S(m+1)$ by taking $m+1$ observations $\mathbf{X}_i$'s, which correspond to smallest $m+1$ $\|r_i(m)\|$'s. Set $m = m + 1$.
5. Iterate $2 - 4$ until $m = n - 1$.
6. The forward plot of the spatial ranks can be obtained by plotting the $r_{min}(m)$ against the corresponding subset sizes m.

This algorithm is computationally easy and straightforward. When the points in $S(m)$ belong to the same cluster, $\|r_i(m)\|$ for a point $\mathbf{X}_i$ belonging to the same cluster is expected to be smaller than that of point from a different cluster. Even if our initial subset contain points from different clusters, the algorithm will ensure that $S(m)$ will move to a single cluster as it grows in size and is constructed by taking points with smallest ranks. So whenever $S(m)$ grows bigger than the cluster it originally belonged to, we expected to see a jump in the magnitude of the rank function as the nearest point to $S(m)$ is then from a different cluster. However, we can observe that $\|rank_F(\mathbf{x})\| < 1$ for all $\mathbf{x} \in \mathbb{R}^d$. Thus, all $\|r_i(m)\|$'s are bounded by 1. Hence, even if a particular point $\mathbf{X}_i$, is far from the cluster $S(m)$, the corresponding $\|r_i(m)\|$ may not be very large compared to an observation $\mathbf{X}_j$, which is an extreme observation in $S(m)$. For this reason, the plot of $r_{min}(m)$ against m may not show any sharp increase even when we include a point from a different cluster, and it becomes visually difficult to detect the clusters. To enhance the visual detection of clusters, we modify the algorithm using central rank regions determined by $r_{min}(m)$. Central rank regions are defined as

$$C_F(r) = \{\mathbf{x} : \|rank_F(\mathbf{x})\| \leq r\}, \qquad 0 < r < 1. \tag{6}$$

One can define the (real-valued) volume functional of the multivariate central ranks region as

$$V_F(r) = volume(C_F(r)), \qquad 0 \leq r < 1. \tag{7}$$

Serfling (2002) pointed out that as an increasing function of r, $V_F(r)$ characterizes the spread of F in terms of expansion of the central regions $C_F(r)$. For each r, $V_F(r)$ is invariant under shift and orthogonal transformations, and $V_F(r)^{1/d}$ is equivariant under homogeneous scale transformations.

We modify Step 6 of the above algorithm and produce a forward plot of the volume functional, $vol(m)$ against the subset size m, where $vol(m)$ is the volume of the central rank region determined by $r_{min}(m)$, i.e., $vol(m) = V_{S(m)}(r_{min}(m))$ based on the subset $S(m)$. Note that, as soon as we include a point from a different cluster in the subset, the volume of the central rank region increases substantially and then it may remain around that large volume as it includes more and more points from that cluster and we may see a sharp decrease in volume after some time if the subset $S(m)$ moves to the new cluster completely. However that depends on the relative cluster sizes and how far they are from each other. Eventually, points from all clusters will be in $S(m)$ and the volume of the central rank regions will grow with m.

In order to compute the volume of the central rank regions, we first compute a discretized boundary of $C_F(r)$ by computing geometric quantiles corresponding to index vector $\mathbf{u}$ with $\|\mathbf{u}\| = r$ following Chaudhuri (1996). Then the volume of the discretized central rank region $C_F(r)$ is computed using the quickhull algorithm of Barber et al. (1996), which was implemented in the R package `geometry`. The computation of volumes may be computationally expensive in very high dimensions. This computational simplification produces an estimate of the volume of $C_F(r)$, however, the precision of the estimate increases with the increase in the number of points chosen on the boundary. We may need to choose the level of discretization sensibly to balance between the computational time and accuracy in estimation. As this is a visualization tool, even if our estimate of volume is not too precise, we are still able to see the distinct jumps for the clusters when they are well separated.

In principle, the initial subset size can be anything more than 1 as the rank of any $\mathbf{x} \in \mathbb{R}^d$ with respect to a single data point is always 1 and we cannot proceed in our algorithm. Also, note that in the modified version of the algorithm, we are computing volumes of central rank regions and as we mentioned earlier that the volume provides a measure of scale, the computation of volumes are meaningful only when the number of observations are at least $d + 1$. Thus, purely for more stability in the algorithm, we choose an initial subset size of $d + 1$. If there are large number of clusters and all are with sizes smaller than $d + 1$, then our algorithm will not be able to estimate the number of clusters efficiently, but that is a rarity for large sample size n.

3 Numerical Examples

We present some systematic evaluations of the proposed forward search algorithms. In the first example, we present the forward searches based on both of spatial ranks and volume of central rank regions on simulated data from three different mixture distributions, namely, multivariate normal, multivariate Laplace and multivariate t with three degrees of freedom for dimensions 2 and 3. Finally, we compare the performance of the forward search based on volume of central rank regions for two different real data sets with two popular clustering methods: mclust approach (Fraley and Raftery 2003) where the best number of groups is chosen according to BIC and k-means where the best number of groups is chosen according to CH index.

3.1 Simulated Data Examples

In the first example, we consider three bivariate mixture distributions with elliptic symmetry. For mixture normal distribution, we take $\mathbf{X}_1, \mathbf{X}_2, \ldots, \mathbf{X}_n \in \mathbb{R}^d$ as a random sample from bivariate mixture normal distribution,

$$p.N_2(\boldsymbol{\mu}_1, \boldsymbol{\Sigma}) + (1-p).N_2(\boldsymbol{\mu}_2, \boldsymbol{\Sigma}), \tag{8}$$

where $\boldsymbol{\mu}_1 = (0,0)^\top$, $\boldsymbol{\mu}_2 = (5,5)^\top$, $\boldsymbol{\Sigma} = \begin{pmatrix} 1 & 0.5 \\ 0.5 & 1 \end{pmatrix}$ and $p = 0.3$. For the second case, we consider multivariate Laplace distribution, $Lap_d(\boldsymbol{\mu}, \boldsymbol{\Sigma})$ with the probability density function,

$$f(\mathbf{x}) \propto \frac{1}{|\boldsymbol{\Sigma}|^{\frac{1}{2}}} e^{-\sqrt{(\mathbf{x}-\boldsymbol{\mu})^\top \boldsymbol{\Sigma}^{-1}(\mathbf{x}-\boldsymbol{\mu})}}, \tag{9}$$

and consider a random sample from the bivariate mixture Laplace distribution,

$$p.Lap_2(\boldsymbol{\mu}_1, \boldsymbol{\Sigma}) + (1-p).Lap_2(\boldsymbol{\mu}_2, \boldsymbol{\Sigma}), \tag{10}$$

with $\boldsymbol{\mu}_1$, $\boldsymbol{\mu}_2$, $\boldsymbol{\Sigma}$ and p as before. For the third case, we consider the multivariate Student's t-distribution with ν degrees of freedom, $t_d(\nu; \boldsymbol{\mu}, \boldsymbol{\Sigma})$ with the probability density function,

$$f(\mathbf{x}) = \frac{\Gamma\left[(\nu+d)/2\right]}{\Gamma(\nu/2)\nu^{d/2}\pi^{d/2}\,|\boldsymbol{\Sigma}|^{1/2}\left[1 + \frac{1}{\nu}(\mathbf{x}-\boldsymbol{\mu})^\top \boldsymbol{\Sigma}^{-1}(\mathbf{x}-\boldsymbol{\mu})\right]^{(\nu+d)/2}}, \tag{11}$$

and consider a random sample from bivariate mixture t-distribution with $\nu = 3$ degrees of freedom,

$$p.t_2(3; \boldsymbol{\mu}_1, \boldsymbol{\Sigma}) + (1-p).t_2(3; \boldsymbol{\mu}_2, \boldsymbol{\Sigma}). \tag{12}$$

In all three cases considered, we generate samples of $n = 100$ observations and produce forward search plots with 100 randomly chosen initial subsets for each as considered in Atkinson and Riani (2007). Our objective is to determine the subsets for the trajectories where there is evidence of a cluster structure. Since our generated data coming from mixture models, with mixture proportions ($p = 0.3$, $1 - p = 0.7$), we expect to get a clearly common structure around subsets with sizes 30 and 70 respectively. Figure 1 is a forward plot of minimum Mahalanobis distances from 100 random starts for samples size $n = 100$ from bivariate mixture normal, Laplace and t distributions with correlated variables. As we can see, only for the normal distribution, there is a common structure around subsets with sizes 30 and 70, respectively. However, the forward plot based on Mahalanobis distance failed to give us a reasonable result for both Laplace and Student's t distributions.

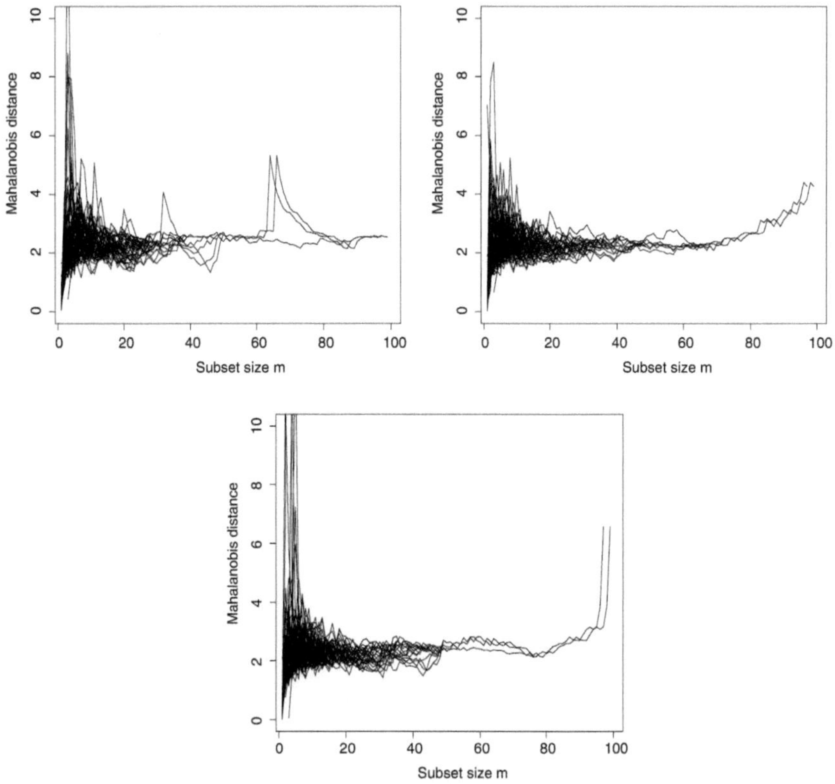

Fig. 1 Forward plot of minimum Mahalanobis distances from 100 randomly chosen initial subsets for sample size $n = 100$ from bivariate mixture normal, Laplace and t distributions (*clockwise from upper left*)

Figure 2 is a forward plot of minimum spatial ranks from 100 random starts for the same simulated data. It can be clearly noticed that there are many different values of $r_{min}(m)$ presented in many trajectories. Moreover, the three plots in Fig. 2 show that there is clearly common structure around subsets with sizes 30 and 70 respectively, where there are two clear maxima in these plots, one at $m = 30$ and the other at $m = 70$, suggesting the existence of two clusters. So these plots lead to the division of the data into two clusters, which means that the forward search based on spatial ranks performs well with the three elliptically symmetric distributions, and it outperforms the one based on Mahalanobis distances for Laplace and t distributions. However, as mentioned earlier, the spatial ranks are bounded by 1 and hence do not produce a good visual effect to detect clusters in an easier way.

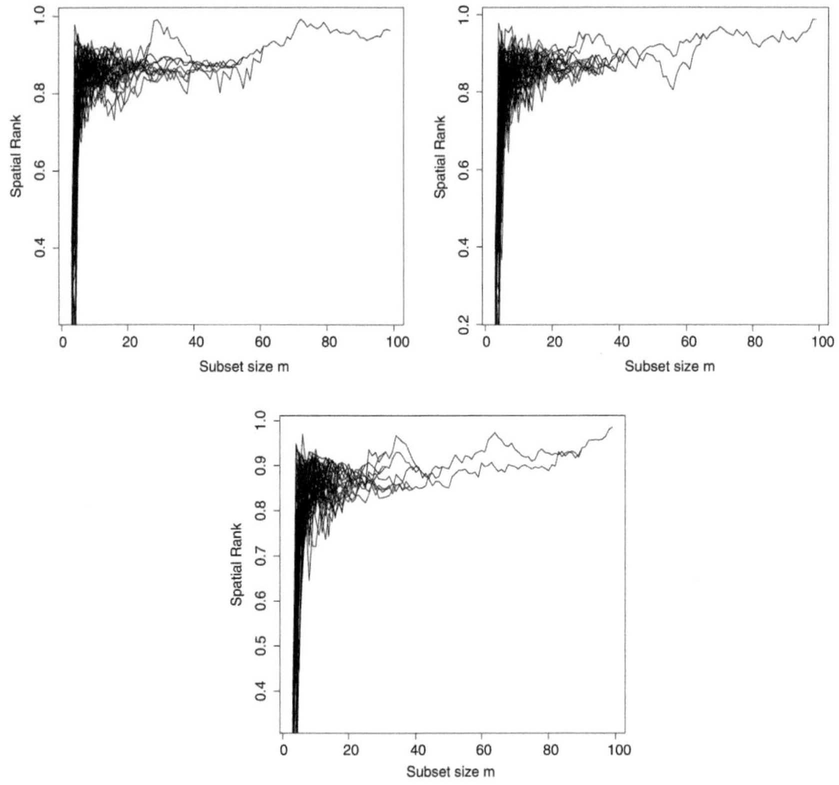

Fig. 2 Forward plot of minimum spatial ranks from 100 randomly chosen initial subsets for sample size $n = 100$ from bivariate mixture normal, Laplace and t distributions (*clockwise from upper left*)

Now, we consider the forward plot based on the volume of central rank regions. Figure 3 is a forward plot of minimum volume functional of central rank regions from 100 random starts for samples size $n = 100$ from bivariate mixture normal, Laplace and t distributions with correlated variables. The three plots in Fig. 3 show that there is clearly a common structure around subsets with sizes 30 and 70 respectively, where there are two clear maxima in these plots, one at $m = 30$ and the other at $m = 70$, suggesting the existence of two clusters. So these plots also lead to the division of the data into two clusters. Compared to the forward search based on Mahalanobis distances and spatial ranks, the forward plot based on volumes of central rank regions gives better results, specially in Laplace and t distributions, where it gives plots with a clearer structure around subsets with sizes 30 and 70. Moreover, it is more accurate in the purpose of visualization since we can easily determine the number of clusters from the plot based on volume of central rank regions. Thus, it should be concluded that the forward search based on volume of central rank regions outperforms forward search based on Mahalanobis distances and spatial ranks.

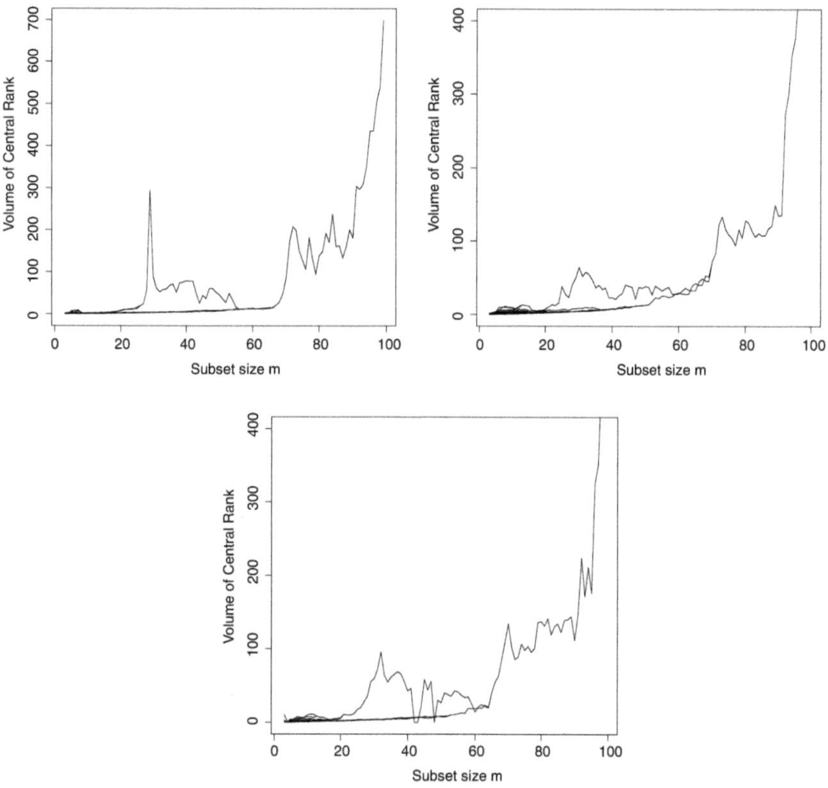

Fig. 3 Forward plot of minimum volume functional of central rank regions from 100 randomly chosen initial subsets for sample size $n = 100$ from bivariate mixture normal, Laplace and t distributions (*clockwise from upper left*)

In the next example, we consider trivariate mixture distributions of normal, Laplace and Student's t with three degrees of freedom, as before with

$$\boldsymbol{\mu}_1 = \begin{pmatrix} 0 \\ 0 \\ 0 \end{pmatrix}, \ \boldsymbol{\mu}_2 = \begin{pmatrix} 5 \\ 5 \\ 5 \end{pmatrix}, \ \boldsymbol{\Sigma} = \begin{pmatrix} 1 & 0.5 & 0.5 \\ 0.5 & 1 & 0.5 \\ 0.5 & 0.5 & 1 \end{pmatrix}. \tag{13}$$

The mixing proportion p is again taken to be 0.3. Figure 4 is a forward plot of minimum Mahalanobis distances. Like Fig. 1, the forward plot based on Mahalanobis distances failed to give us a common structure around subsets with sizes 30 and 70 in Laplace and Student's t distributions. The only reasonable result was for the trivariate mixture normal distribution.

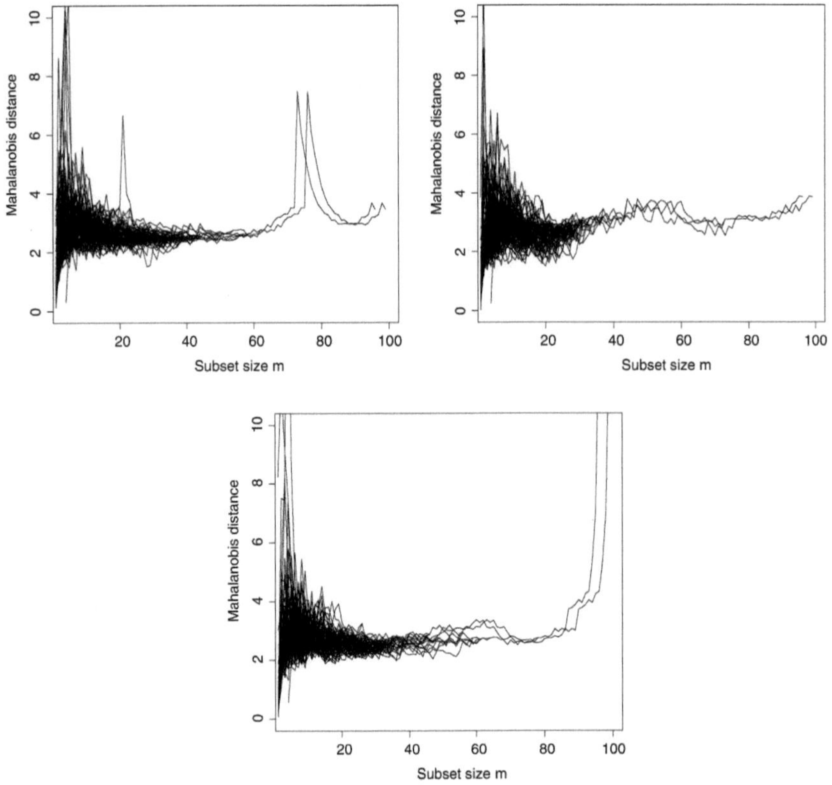

Fig. 4 Forward plot of minimum Mahalanobis distances from 100 randomly chosen initial subsets for sample size $n = 100$ from trivariate mixture normal, Laplace and t distributions (*clockwise from upper left*)

Figure 5 is a forward plot of minimum spatial ranks. There are two clear maxima one around $m = 30$ and the other around $m = 70$, which can be considered as indicator of the existence of two clusters. Compared to Fig. 4, we can see that Fig. 5 gives better results, where it gives plots with a clearer structure around the subsets with sizes 30 and 70, which means that the forward search based on spatial ranks gives better result for the data with higher dimensions. Moreover, it outperforms the one based on Mahalanobis distances for Laplace and t distributions.

On the other hand, for the forward search based on volume of central rank regions, we can notice from Fig. 6 that the performance has been also improved, where it gives two very clear maxima at $m = 30$ and at $m = 70$ and better than Fig. 5 as well.

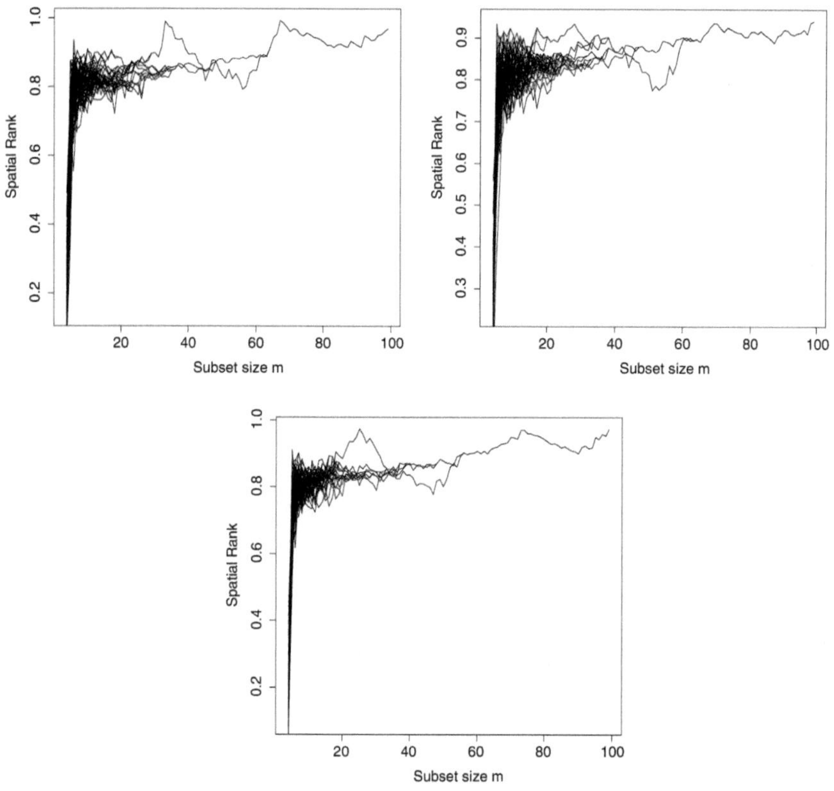

Fig. 5 Forward plot of minimum spatial ranks from 100 randomly chosen initial subsets for sample size $n = 100$ from trivariate mixture normal, Laplace and t distributions (*clockwise from upper left*)

In Fig. 7, we present an example of a forward plot based on volumes of central rank regions, where the data is simulated from a mixture of 3 bivariate normal distributions,

$$p_1 N_2(\boldsymbol{\mu}_1, \boldsymbol{I}) + p_2 N_2(\boldsymbol{\mu}_2, \boldsymbol{I}) + (1 - p_1 - p_2) N_2(\boldsymbol{\mu}_3, \boldsymbol{I}),$$

where $\boldsymbol{\mu}_1 = (0, 4)^\top, \boldsymbol{\mu}_2 = (-4, -4)^\top, \boldsymbol{\mu}_3 = (4, -4)^\top$ and $p_1 = 0.2, p_2 = 0.3$. With trajectories from 100 randomly chosen initial subsets, we see a clear pattern of three cluster sizes here.

3.2 Real Data Examples

The first dataset we consider is known as the Old Faithful Geyser Data, which are taken from Azzalini and Bowman (1990) and the `MASS` library Venables and Ripley (2002). This data gives the waiting time between eruptions and the duration of

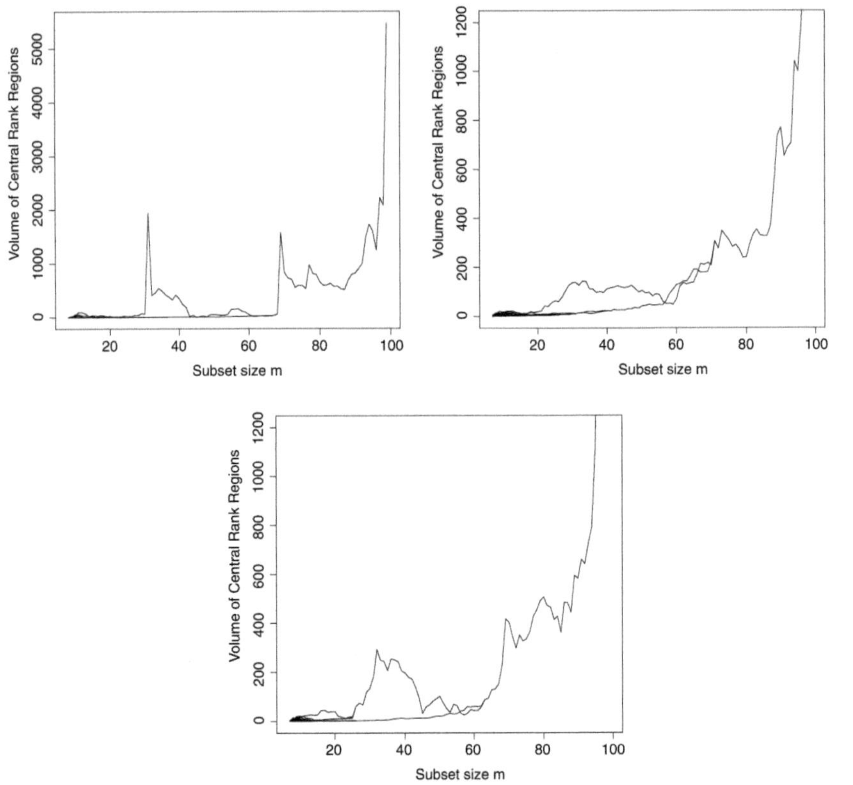

Fig. 6 Forward plot of minimum volume functional of central rank regions from 100 randomly chosen initial subsets for sample size $n = 100$ from trivariate mixture normal, Laplace and t distributions (*clockwise from upper left*)

the eruption in minutes for the Old Faithful geyser in Yellowstone National Park, Wyoming, USA, with two apparent groups in the data. The analysis of this data using the standard forward approach based on Mahalanobis distances had been done in Atkinson and Riani (2012). It includes 272 observations with two variables, x_{1i}: the duration of the ith eruption and x_{2i}: the waiting time to the start of that eruption from the start of eruption $i - 1$. For the k-means, the selection criterion that we use is the CH-index (Calinski and Harabasz 1974), where we use it to estimate the number of clusters that k-means algorithm should start with it, and for the BIC criterion we use the `mclust` library (Fraley and Raftery 2003), where Fraley and Raftery (2003) assumed 10 models of the parameterization of the Gaussian mixture models

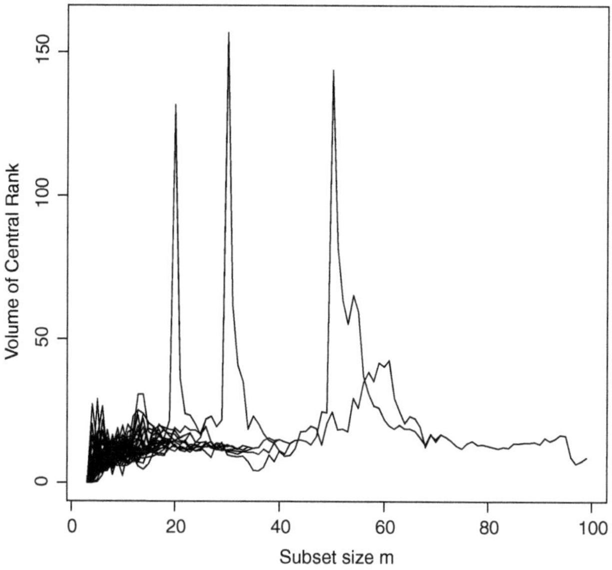

Fig. 7 Forward plot based on volumes of central rank regions with 100 randomly chosen initial subsets for a mixture bivariate normal data set with 3 mixing densities

(parsimonious models) introduced earlier by Banfield and Raftery (1993). Figure 8 shows the behavior of the CH-index, k-means, BIC, and forward search with volume of central rank regions. The upper left and right panels of Fig. 8 are the CH-index plot which indicates ten clusters and the clustering with k-means respectively. Clearly, the k-means behaves so poorly in this real dataset, where it failed to give us the right clustering. On the other hand, from the lower left panel we can see that the best model according to BIC is an equal-covariance model with three clusters, where the maximum value of the BIC criterion among the 10 parsimonious models was for the EEE model (BIC $= -2314.386$) with similarly shaped covariance matrices, and the next best model had four clusters (BIC $= -2320.207$) with the same covariance structure. This indicates that the mclust approach based on BIC criterion failed to give the right number of clusters as well as k-means method. For our methodology, the lower right panel of Fig. 8 shows the forward plot of volume functional of central rank regions among units not in the subset from 100 random starts for Old Faithful data. There are two clear maxima in this plot, one at $m = 105$ and the other at $m = 179$, suggesting the existence of two clusters. This in fact considers a better result compared to k-means and mclust approach.

The second real dataset used in this article, is a financial data contains measurements on three variables monitoring the performance of 103 investment funds operating in Italy since April 1996 [Table A.16 of Atkinson et al. (2004)]. These three variables are, y_1: short term (12 month) performance, y_2: medium term (36 month) performance, and y_3: medium term (36 month) volatility. Additionally, this data

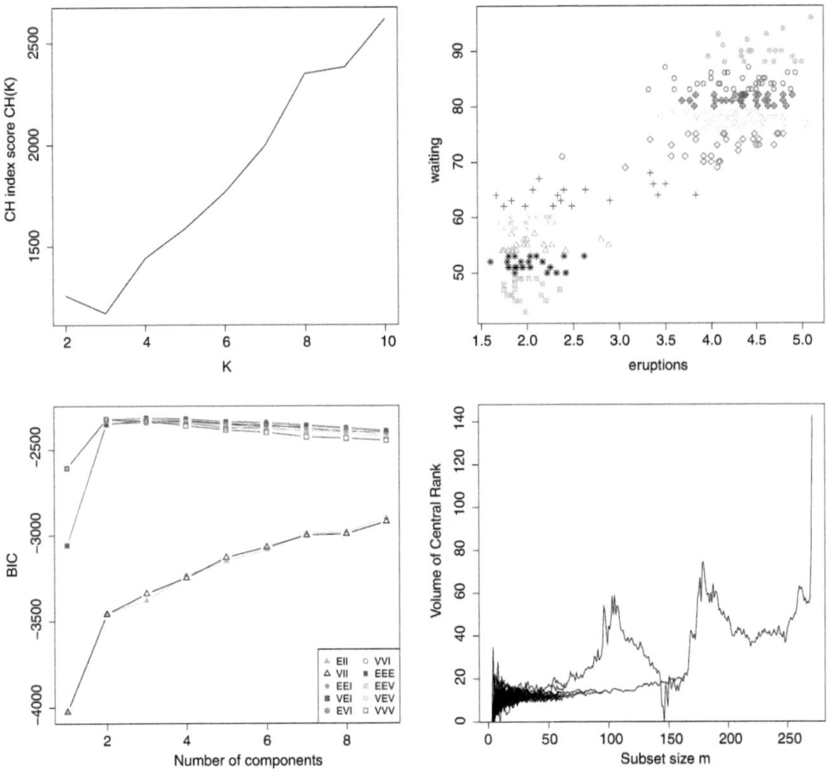

Fig. 8 Old Faithful data: *clockwise from upper left*: CH index suggests k = 10, k-means with 10 clusters, BIC plot suggesting 3 clusters with best BIC values for EEE model, and forward plot of volumes of central rank regions with 100 randomly chosen initial subsets; two clusters are evident at $m = 105$ and 179

include two different kinds of fund, since the units 1–56 are all stock funds whereas units 57–103 are balanced funds. Atkinson et al. (2004) and Atkinson et al. (2006) applied their forward search method based on Mahalanobis distances to cluster these financial data and introduced detailed analysis of it. According to Fig. 9, like our method, k-means indicated two clusters, while the mclust approach based on BIC again failed to give the true number of clusters, where the maximum value of the BIC criterion was for the EEE model (BIC $= -1664.278$). The lower right panel of Fig. 9 is a forward plot based on volume of central rank regions among units not in the subset. There are two clear maxima again one at $m = 44$ and the other at $m = 56$ which leads to the division of the data into two clusters showing the successful clustering of our method.

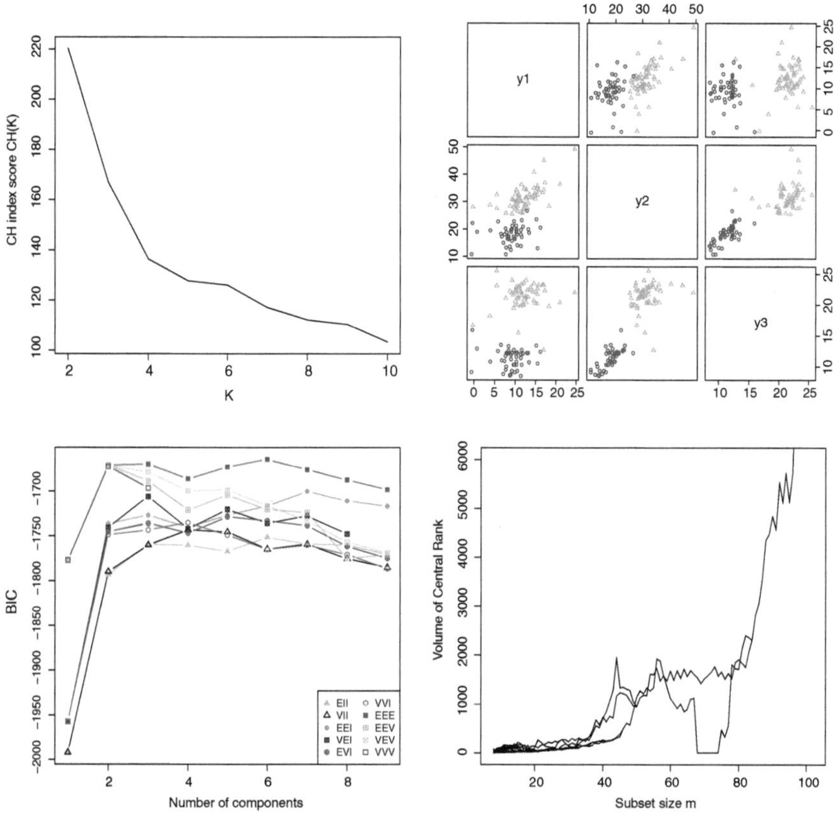

Fig. 9 Financial data: *clockwise from upper left*: CH index suggests k = 2, k-means with two clusters, BIC plot suggesting six clusters with best BIC values for EEE model, and forward plot of volumes of central rank regions with 100 randomly chosen initial subsets; two clusters are evident at $m = 44$ and 56

4 Concluding Remarks

A forward search algorithm is considered in this article, which is based on nonparametric multivariate spatial rank functions and it is robust in terms of determining the number of clusters by the data itself. Using nonparametric approaches helps to get techniques which are less sensitive to the statistical model assumptions, and solve problems such as the heavy tailed distributed data with high level of correlation among the variables. The forward search plots based on spatial ranks show that the algorithm performs well under heavy tailed mixture distributions with elliptic symmetry and it outperforms the forward search based on Mahalanobis distances for nonnormal mixture distributions. The modified forward search plots based on volume of central rank regions outperforms the forward search based on Mahalanobis distances and spatial ranks as illustrated in the numerical examples. More visually clear results have been obtained using the volume of central rank regions, since it gives forward plots with a clearer structure of clusters.

In all of numerical examples, the mixture densities $f_1, \ldots, f_k$ are from the same family of distributions. We should mention that it is not necessary to assume them to be coming from the same parametric family as we are not estimating any parameters in our proposed visual tool. This is one of the greatest advantages of the proposed method. However for large number of clusters, the proposed forward search plots may produce too many peaks and makes it very difficult visually to determine the number of clusters and the cluster sizes. At present, we do not have any formal procedure to estimate the number of clusters from this plot, and we are looking into that problem as a future research project. With a formal procedure, we should able to validate the estimate against the model assumptions.

It is well known that spatial ranks are invariant under orthogonal transformations, but they are not invariant under general affine transformations of the data and hence the proposed algorithms are not affine invariant. To make the algorithms affine invariant, one may look for affine invariant versions of spatial ranks (see for example, Chakraborty 2001) and follow the same algorithm to construct the forward search plot. To keep the simplicity of the algorithms and to save on computational time, we refrained from using affine invariant versions of spatial ranks in this work. Using affine invariant ranks may improve the results if the scales of different clusters are not similar.

Acknowledgments The authors would like to greatly thank the editors of ICORS 2015 and the two referees for their helpful remarks and comments on an earlier version of the manuscript. The research of Mohammed Baragilly is partially supported by the Egyptian Government and he would like to express his greatest appreciation to the Egyptian Cultural Centre and Educational Bureau in London and to the Department of Applied Statistics, Helwan University.

References

Atkinson AC (1994) Fast very robust methods for the detection of multiple outliers. J Am Stat Assoc 89:1329–1339

Atkinson AC, Mulira H (1993) The stalactite plot for the detection of multivariate outliers. Stat Comput 3:27–35

Atkinson AC, Riani M (2007) Exploratory tools for clustering multivariate data. Comput Stat Data Anal 52:272–285

Atkinson AC, Riani M (2012) Discussion on the paper by spiegelhalter, sherlaw-johnson, bardsley, blunt, wood and grigg. J Roy Stat Soc 175

Atkinson AC, Riani M, Cerioli A (2004) Exploring multivariate data with the forward search. Springer, NewYork

Atkinson AC, Riani M, Cerioli A (2006) Random start forward searches with envelopes for detecting clusters in multivariate data. Springer, Berlin, pp 163–171

Atkinson AC, Riani M, Cerioli A (2010) The forward search: theory and data analysis. J Korean Stat Soc 39:117–134

Azzalini A, Bowman A (1990) A look at some data on the old faithful geyser. J Roy Stat Soc 39(3):357–365

Banfield J, Raftery AE (1993) Model-based gaussian and non-gaussian clustering. Biometrics 49:803–821
Barber CB, Dobkin DP, Huhdanpaa H (1996) The quickhull algorithm for convex hulls. ACM Trans Math Softw 22(4):469–483
Beale EML (1969) Euclidean cluster analysis. ISI, Voorburg, Netherlands
Calinski RB, Harabasz J (1974) A dendrite method for cluster analysis. Commun Stat 3:1–27
Chakraborty B (2001) On affine equivariant multivariate quantiles. Ann Inst Stat Math 53:380–403
Chaudhuri P (1996) On a geometric notion of multivariate data. J Am Stat Assoc 90:862–872
Duda RO, Hart PE (1973) Pattern classification and scene analysis. Wiley, New York
Everitt B, Landau S, Leese M, Stahl D (2011) Cluster analysis, 5th edn. Wiley, Chichester
Fraley C, Raftery A (2003) Enhanced model-based clustering, density estimation and discriminant analysis: Mclust. J Classif 20(263):286
Friedman HP, Rubin J (1967) On some invariant criteria for grouping data. J Am Stat Assoc 62:1159–1178
Gan G, Ma C, Wu J (2007) Data clustering theory, algorithms, and applications. ASA-SIAM series on statistics and applied probability. Philadelphia
Gordon AD (1998) Cluster validation. In: C Hayashi KYeae, N Ohsumi (eds) Data science, classification and related methods. Springer, Tokyo, pp 22–39
Hadi AS (1992) Identifying multiple outliers in multivariate data. J Roy Stat Soc 54:761–771
Hadi AS, Simonoff JS (1993) Procedures for the identification of multiple outliers in linear models. J Am Stat Assoc 88(424):1264–1272
Hartigan JA (1975) Clustering algorithms. Wiley, New York
Kaufman L, Rousseeuw PJ (1990) Finding groups in data. Wiley, New York
Koltchinskii V (1997) M-estimation, convexity and quantiles. Ann Stat 25:435–477
Krzanowski WJ, Lai YT (1985) A criterion for determining the number of clusters in a data set. Biometrics 44(23):34
Marriott FHC (1971) Practical problems in a method of cluster analysis. Biometrics 27:501–514
Milligan GW, Cooper MC (1985) An examination of procedures for determining the number of clusters in a data set. Psychometrika 50:159–179
Mojena R (1977) Hierarchical grouping methods and stopping rules: an evaluation. Comput J 20:359–363
Overall JE, Magee KN (1992) Replication as a rule for determining the number of clusters in hierarchical cluster analysis. Appl Psychol Measur 16:119–128
Serfling R (2002) A depth function and a scale curve based on spatial quantiles. In: Dodge Y (ed) Statistical data analysis based on the L1-norm and related methods. Birkhaeuser, pp 25–38
Sugar CA, James GM (2003) Finding the number of clusters in a data set: an information theoretic approach. J Am Stat Assoc 98:750–763
Thorndike RL (1953) Who belongs in a family? Psychometrika 18:267–276
Tibshirani R, Walther G, Hastie T (2001) Estimating the number of clusters in a data set via the gap statistic. J Roy Stat Soc 63:411–423
Venables W, Ripley B (2002) Modern applied statistics with S, 4th edn. Springer, NewYork

Robust and Sparse Estimation of the Inverse Covariance Matrix Using Rank Correlation Measures

Christophe Croux and Viktoria Öllerer

1 Introduction

We have a sample of n multivariate observations, and for each of these observations we measure p variables. The resulting data can be collected in a data matrix $\mathbf{X}$ where the observations are the rows of the data matrix, and each variable corresponds to a column of the data matrix. The data matrix $\mathbf{X}$ has np cells, where a cell contains a univariate measurement x_{ij}

$$\mathbf{X} = \begin{pmatrix} x_{11} & \ldots & \ldots & x_{1p} \\ \vdots & \vdots & \vdots & \vdots \\ x_{n1} & \ldots & \ldots & x_{np} \end{pmatrix}.$$

Typically, these data matrices are thin, with n much larger than p. But in this paper focus is on fat data matrices with more columns than rows. Fat data matrices often occur in practice. For instance in medicine where hundreds of variables are measured for a limited set of patients. The transposed rows of $\mathbf{X}$ are denoted as $\mathbf{x}_1, \ldots, \mathbf{x}_n$, with $\mathbf{x}_i \in \mathbb{R}^p$. The columns of the data matrix are denoted as $\mathbf{x}^1, \ldots, \mathbf{x}^p \in \mathbb{R}^n$.

We assume that the observations are a random sample of a multivariate normal distribution with mean $\boldsymbol{\mu}$ and covariance matrix $\boldsymbol{\Sigma}$. This covariance matrix is assumed to be positive definite, hence, all its eigenvalues are strictly positive. The aim is to estimate the unknown parameters $\boldsymbol{\mu}$ and $\boldsymbol{\Sigma}$ from the data such that (i) the estimators are resistant to outlying cells (ii) the estimate of $\boldsymbol{\Sigma}$ is positive definite.

In high dimensions, the occurrence of outliers is to be expected. Data are collected less carefully, often in an automatic and inaccurate way. Gross-errors can occur.

C. Croux (✉) · V. Öllerer
ORSTAT, Faculty of Economics and Business, KU Leuven, 3000 Leuven, Belgium
e-mail: christophe.croux@kuleuven.be

V. Öllerer
e-mail: viktoria.oellerer@kuleuven.be

C. Agostinelli et al. (eds.), *Recent Advances in Robust Statistics: Theory and Applications*, DOI 10.1007/978-81-322-3643-6_3

Moreover, the size of the data set and the large number of variables makes outlier detection using visualization cumbersome. Therefore, estimators should be robust to outlying values x_{ij}, hence outlying cells. In the traditional literature on robust statistics (see Maronna et al. 2006, for a more recent textbook), one considers outlying observations, and an observation is already an outlier if only one of its cells is outlying. In high dimensions, the notion of outlying cells is more appropriate. Indeed, take $p = 200 > n = 100$ and assume that every cell x_{ii}, for $1 \leq i \leq n$ is an outlier. Then all observations are outliers, suggesting that robust estimation would be impossible. But only 0.5 % of the cells are outliers. This contamination model, as well as more general settings appropriate for high-dimensional data, are introduced in Alqallaf et al. (2009). They also discuss challenges in high-dimensional robust analysis and take steps to measure robustness in these general contamination settings. In Sect. 4 we define the concept of breakdown point under cellwise contamination, as introduced in Öllerer and Croux (2015). The estimators advocated in this paper have a high breakdown point according to this definition, showing that robust estimators do exist in high dimensions. One only needs to reconsider what appropriate measures for robustness are in high dimensions.

The sample covariance matrix estimator

$$\hat{\boldsymbol{\Sigma}} = \frac{1}{n-1}\sum_{i=1}^{n}(\mathbf{x}_i - \bar{\mathbf{x}})(\mathbf{x}_i - \bar{\mathbf{x}})^\top \tag{1}$$

with $\bar{\mathbf{x}}$ the sample average is not only nonrobust, it also has the problem that it is only positive semidefinite. Some of its eigenvalues will be zero if $p \geq n$. Hence, its inverse is not existing. In multivariate statistics one often needs the inverse: to compute Mahalanobis distances, for Fisher discriminant analysis, ... Therefore, we want to have an estimator of $\boldsymbol{\Sigma}$ that is always positive definite. We can achieve this in many ways, but a popular choice is the Graphical Lasso, or Glasso, of Friedman et al. (2008). Glasso takes a positive semidefinite covariance matrix estimator as an input, and returns a positive definite one. A particular feature of Glasso is that the resulting estimator of the inverse covariance matrix is also sparse, meaning that many of its element are exactly equal to zero. We denote the inverse covariance matrix, or *precision matrix*, as $\boldsymbol{\Theta} = \boldsymbol{\Sigma}^{-1}$. The classical Glasso estimator takes the nonrobust sample covariance matrix as an input, thus, it is only suitable for clean data that do not contain any outliers. Therefore, we will replace the sample covariance matrix with a robust covariance estimate and show that this indeed leads to a robust precision matrix estimate.

In Sect. 2, we define the estimators of the precision matrix to be studied. They are robust to cellwise outliers, and give sparse and positive definite estimates of $\boldsymbol{\Theta}$. In Sect. 3, we give some R code to show how easily the estimates can be computed. Theoretical results are presented in Sect. 4. The different estimators are compared in Sect. 5 by means of a simulation experiment. Section 6 shows how the estimators have been used for graphical modeling. Section 7 contains some final discussion.

2 Estimators

We follow the approach of Tarr et al. (2016) for constructing sparse and robust precision matrices. In a first step, we construct a robust estimator $\mathbf{S}$ of the covariance matrix. In a last step, $\mathbf{S}$ serves as an input of Glasso, resulting in a sparse and robust estimator $\hat{\boldsymbol{\Theta}}_{\mathbf{S}}$ of the precision matrix

$$\hat{\boldsymbol{\Theta}}_{\mathbf{S}} = \arg\max_{\substack{\boldsymbol{\Theta}=(\theta_{jk})\in\mathbb{R}^{p\times p}\\ \boldsymbol{\Theta}\succ 0}} \log\det(\boldsymbol{\Theta}) - \operatorname{tr}(\mathbf{S}\boldsymbol{\Theta}) - \lambda\sum_{j,k=1}^{p}|\theta_{jk}|, \tag{2}$$

where the maximization is over all positive definite matrices $\boldsymbol{\Theta}\succ 0$. The algorithm for solving (2) requires the input matrix $\mathbf{S}$ to be symmetric and positive semidefinite. A stable implementation of Glasso is given in the R-package `huge` (Zhao et al. 2014a). The parameter λ in (2) controls for the sparsity of the solution: the larger λ, the sparser the precision matrix estimate. We compute $\hat{\boldsymbol{\Theta}}_{\mathbf{S}}$ over a logarithmic spaced grid of ten values, as is done by default in the `huge`-package. The final solution is then the one with lowest value of the following Bayesian Information Criterion (see Yuan and Lin 2007)

$$BIC(\lambda) = -\log\det\hat{\boldsymbol{\Theta}}_{\mathbf{S}} + \operatorname{tr}(\hat{\boldsymbol{\Theta}}_{\mathbf{S}}\mathbf{S}) + \frac{\log n}{n}\sum_{i\le j}\hat{e}_{ij}(\lambda). \tag{3}$$

with $\hat{e}_{ij}=1$ if $(\hat{\boldsymbol{\Theta}}_{\mathbf{S}})_{ij}\neq 0$ and $\hat{e}_{ij}=0$ otherwise. Note that $\hat{\boldsymbol{\Theta}}_{\mathbf{S}}$ depends on λ.

2.1 Two-Step Estimators

So how do we choose $\mathbf{S}$? Tarr et al. (2016) propose to use the robust covariance of Gnanadesikan and Kettenring (1972) between $\mathbf{x}^j$ and $\mathbf{x}^k$ for s_{jk}, with s_{jk} an element of $\mathbf{S}$. Öllerer and Croux (2015) showed that this choice leads to some loss of robustness and a too high computational cost. Instead they propose to use

$$s_{jk} = \operatorname{scale}(\mathbf{x}^j)\operatorname{scale}(\mathbf{x}^k) r(\mathbf{x}^j,\mathbf{x}^k) \qquad j,k=1,\dots,p. \tag{4}$$

As scale estimator scale() the robust Q_n-estimator (Rousseeuw and Croux 1993) is taken, which has the highest possible breakdown point of all scale estimator and is quite efficient at the normal model. For the correlation $r(\mathbf{x}^j,\mathbf{x}^k)$ Öllerer and Croux (2015) considered the following three choices:

- The Quadrant correlation, defined as

$$r_{Quadrant}(\mathbf{x}^j,\mathbf{x}^k) = \frac{1}{n}\sum_{i=1}^{n}\operatorname{sign}((x_{ij}-\operatorname{med}_{\ell=1,\dots,n}x_{\ell j})(x_{ik}-\operatorname{med}_{\ell=1,\dots,n}x_{\ell k})), \tag{5}$$

where sign(·) denotes the sign-function. The use of Quadrant correlation was advocated in Alqallaf et al. (2002).

- The Spearman correlation defined as the sample correlation of the ranks of the observations

$$r_{Spearman}(\mathbf{x}^j, \mathbf{x}^k) = \sum_{i=1}^{n} \frac{(R(x_{ij}) - \frac{n+1}{2})(R(x_{ik}) - \frac{n+1}{2})}{\sqrt{\sum_{i=1}^{n}(R(x_{ij}) - \frac{n+1}{2})^2 \sum_{i=1}^{n}(R(x_{ik}) - \frac{n+1}{2})^2}}, \quad (6)$$

with $R(x_{ij})$ the rank of x_{ij} among all elements of $\mathbf{x}^j$, for any $1 \leq j \leq p$ and $1 \leq i \leq n$.

- The Gaussian rank correlation defined as the sample correlation estimated from the normal scores of the data

$$r_{Gauss}(\mathbf{x}^j, \mathbf{x}^k) = \frac{\sum_{i=1}^{n} \Phi^{-1}(\frac{R(x_{ij})}{n+1})\Phi^{-1}(\frac{R(x_{ik})}{n+1})}{\sum_{i=1}^{n}(\Phi^{-1}(\frac{i}{n+1}))^2}, \quad (7)$$

where $\Phi(\cdot)$ is the cumulative distribution function of a standard normal.

In this paper, we also consider a fourth correlation measure.

- Kendall correlation (Kendall 1938) is based on signs

$$r_{Kendall}(\mathbf{x}^j, \mathbf{x}^k) = \frac{2}{n(n-1)} \sum_{1 < i \leq i' < n} \operatorname{sign}((x_{ij} - x_{i'j})(x_{ik} - x_{i'k})). \quad (8)$$

Its drawback is the slow computation. Even though there exists a $O(n \log n)$ algorithm (implemented as `cor.fk()` in the R-package `pcaPP` (Todorov et al. 2014)), it is somewhat slower than the computation of Spearman correlation.

The robustness and efficiency properties of the Quadrant, Kendall, and Spearman correlation are studied in Croux and Dehon (2010), and of the Gaussian rank correlation in Boudt et al. (2012). Using these correlation measures, combined with (4), yields positive semidefinite covariance matrices.

2.2 Three-Step Estimators

The Quadrant, Kendall, and Spearman correlation are not consistent at the bivariate normal distribution. This means that Quadrant, Kendall, and Spearman correlation between two variables having a joint normal distribution with correlation ρ do not estimate ρ, not even if the sample size is infinite. The corresponding $\mathbf{S}$ is not a consistent estimator of $\boldsymbol{\Sigma}$ and has an asymptotic bias. To resolve this inconsistency, the following transformations need to be applied:

$$\tilde{r}_{Quadrant} = \sin(\frac{\pi}{2} r_{Quadrant}), \quad (9)$$

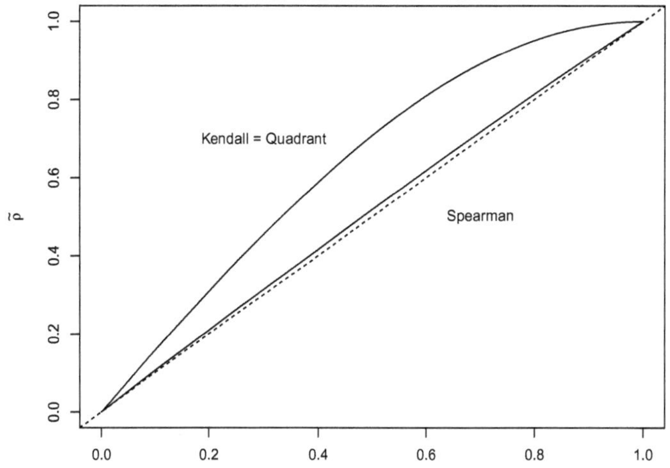

Fig. 1 Plot of the transformations $\rho \to \sin(\pi\rho/2)$ and $\rho \to 2\sin(\pi\rho/6)$ needed for making Quadrant/Kendall and Spearman correlation consistent, together with the 45° line

$$\tilde{r}_{Kendall} = \sin(\frac{\pi}{2} r_{Kendall}), \tag{10}$$

$$\tilde{r}_{Spearman} = 2\sin(\frac{\pi}{6} r_{Spearman}). \tag{11}$$

Hence, to get consistency, the transformed Spearman, Kendall, and Quadrant correlation need to be plugged into (4). It is instructive to plot the transformations (9), (10) and (11). We see from Fig. 1 that the asymptotic bias of the Spearman correlation is very small; the transformation pushes the Spearman correlation only slightly upwards. On the other hand, the Quadrant and Kendall correlations are more severely underestimating the population correlation ρ.

Unfortunately, the resulting $\mathbf{S}$ will not be positive semidefinite anymore, and cannot be used safely as input for Glasso. Therefore, an additional step to make $\mathbf{S}$ positive semidefinite is needed before Glasso can be applied. We implement two easy ways to do this, but other possibilities do exist (see Zhao et al. 2014b). Denote λ_j and $\mathbf{v}_j$ the eigenvalues and eigenvectors of the matrix $\mathbf{S}$, respectively, for $1 \leq j \leq p$. Since $\mathbf{S}$ is symmetric, these eigenvalues exist as real numbers, but may be negative.

1. The perturbation method is an heuristic approach often used in regularization. One simply adds a nonnegative value to all diagonal elements of $\mathbf{S}$

$$\mathbf{S}_{perturb} = \mathbf{S} + |\min(0, \min_j \lambda_j)|\,\mathbf{I}. \tag{12}$$

It is immediate to see that the resulting covariance matrix has no negative eigenvalues any more.

2. Rousseeuw and Molenberghs (1993) proposed to use

$$\mathbf{S}_{npd} = \sum_{j=1}^{p} \max(0, \lambda_j)\mathbf{v}_j\mathbf{v}_j^t. \tag{13}$$

It has been shown (e.g., Zhao et al. 2014b) that $\mathbf{S}_{npd}$ is the positive semidefinite matrix *nearest* to $\mathbf{S}$, where nearness is measured with the Frobenius matrix norm. Hence the abbreviation *npd*, nearest positive (semi)definite matrix.

So three steps are needed: (i) compute $\mathbf{S}$ (ii) make it a positive semidefinite matrix using (12) or (13) (iii) compute Glasso using the step two matrix as input. These three steps have been used in Tarr et al. (2016) as well, using the npd algorithm of Higham (2002) and a different choice of $\mathbf{S}$. An advantage of the Gaussian rank correlation (7) is that it is already consistent at the normal distribution, without any additional transformation needed. Then a two-step approach is sufficient.

3 Computation

In this section, we show how easily the sparse and robust precision matrix estimators can be computed in the software package R. In case an estimate of $\boldsymbol{\Sigma}$ is needed, one simply needs to invert the estimated precision matrix. The function below implements the Eqs. (12) and (13); the input is a symmetric matrix `sigma` the output a positive semidefinite matrix `sigma.psd`.

```
easy.psd <- function(sigma,method="perturb") {
  if (method=="perturb") {
    p <- ncol(sigma)
    eig <- eigen(sigma, symmetric=TRUE, only.values=TRUE)
    const <- abs(min(eig$values,0))
    sigma.psd <- sigma+diag(p)*const
  }
  if (method=="npd") {
    eig <- eigen(sigma, symmetric=TRUE)
    d <- pmax(eig$values,0)
    sigma.psd <- eig$vectors%*%diag(d)%*%t(eig$vectors)
  }
  return(sigma.psd)
}
```

Assume that the data matrix is in the matrix object x. The positive semidefinite matrix **S** based on the transformed Quadrant correlation is computed by the function below

```
quadrant.transformed <- function(x,method="perturb") {
  x.m <- apply(x,2,median)
  x <- sweep(x,2,x.m)
  x.s <- sign(x)
  x.q <- apply(x,2,Qn)
  cor.quadrant <- sin(pi*cor(x.s)/2)
  sigma.quadrant <- diag(x.q)%*%cor.quadrant%*%diag(x.q)
  return(easy.psd(sigma.quadrant,method))
}
```

To compute the Q_n scale estimator, the R-package robustbase (Rousseeuw et al. 2015) is needed. For the transformed Spearman correlation we get the corresponding **S** as

```
spearman.transformed <- function(x,method="perturb") {
  x.r <- apply(x,2,rank)
  x.q <- apply(x,2,Qn)
  cor.sp <- 2*sin(pi*cor(x.r)/6)
  sigma.sp <- diag(x.q)%*%cor.sp%*%diag(x.q)
  return(easy.psd(sigma.sp,method))
}
```

The following function computes the positive semidefinite matrix **S** based on the transformed Kendall correlation:

```
kendall.transformed <- function(x,method="perturb") {
  x.q <- apply(x,2,Qn)
  cor.kendall <- sin(pi*cor.fk(x)/2) # cor.fk(), package pcaPP
  sigma.kendall <- diag(x.q)%*%cor.kendall%*%diag(x.q)
  return(easy.psd(sigma.kendall,method))
}
```

The covariance matrix from the Gaussian rank correlations (7) is computed by the function

```
Grank <- function(x) {
  n <- nrow(x)
  x.q <- apply(x,2,Qn)
  x.r <- apply(x,2,rank)
  cor.Grank <- cor(qnorm(x.r/(n+1)))
  sigma.Grank <- diag(x.q)%*%cor.Grank%*%diag(x.q)
  return(sigma.Grank)
}
```

where we recall that no transformation is needed.

The final step is to compute Glasso, with sparsity parameter λ selected by minimizing the BIC criterion (3). The huge package of Zhao et al. (2012) allows to do this conveniently. The input of the function below is a positive semidefinite matrix sigma.psd, and the output a sparse precision matrix estimate.

```
theta.sparse <- function(sigma.psd,n) {
  huge.out <- huge(sigma.psd,method="glasso",verbose=FALSE)
  my.bic <- -huge.out$loglik+huge.out$df*log(n)/n
  opt.i <- which.min(my.bic)
  return(huge.out$icov[[opt.i]])
}
```

For the approach based on the Spearman correlations, for instance, and given a data matrix x, the next lines compute the positive semidefinite covariance matrix estimator $\mathbf{S}_{npd}$ and the corresponding precision matrix $\boldsymbol{\Theta}_{\mathbf{S}_{npd}}$.

```
S.hat <- spearman.transformed(x,method="npd")
Theta <- theta.sparse(S.hat,n=nrow(x))
```

Table 1 presents computation times for samples of size $n = 50$, averaged over $M = 10$ simulation runs and over the different sampling distributions used in the Simulation Sect. 5. Comparing the two-step and three-step estimators, one sees that there is only a marginal increase in computation time. Comparing the perturbation method (12) and the nearest positive definite approach (13) one sees that the perturbation method is faster, but the relative difference is marginal. Kendall is as expected

Table 1 Computation time (in seconds) with $n = 50$ averaged over $M = 10$ and over all simulation schemes

	$p = 3$	$p = 30$	$p = 100$
2-step Quadrant	1.24	1.22	1.29
3-step Quadrant (npd)	1.24	1.23	1.28
3-step Quadrant (perturb)	1.25	1.22	1.27
2-step Spearman	1.23	1.23	1.27
3-step Spearman (npd)	1.24	1.22	1.28
3-step Spearman (perturb)	1.23	1.22	1.25
2-step Kendall	1.22	1.23	1.35
3-step Kendall (npd)	1.24	1.23	1.40
3-step Kendall (perturb)	1.23	1.20	1.29
3-step Spatial Sign (npd)	1.24	1.31	2.67
3-step Spatial Sign (perturb)	1.21	1.32	2.67
2-step Gaussian Rank	1.24	1.23	1.29
Glasso	1.28	1.27	1.30

a bit slower than Spearman and Quadrant. But all computation times in Table 1 are relatively close to each other, showing that almost all computation time is taken by computing the Glasso in (2). Only the spatial sign correlation, to be explained in Sect. 5, is considerably slower than the other approaches.

4 Breakdown Point

A definition of breakdown point appropriate for measuring robustness of high-dimensional precision matrices is given in Öllerer and Croux (2015). Define for any symmetric $p \times p$ matrices $\mathbf{A}$ and $\mathbf{B}$

$$D(\mathbf{A}, \mathbf{B}) = \max\{|\lambda_1(\mathbf{A}) - \lambda_1(\mathbf{B})|, |\lambda_p(\mathbf{A})^{-1} - \lambda_p(\mathbf{B})^{-1}|\},$$

where the ordered eigenvalues of a matrix $\mathbf{A}$ are denoted by $0 \leq \lambda_p(\mathbf{A}) \leq \ldots \leq \lambda_1(\mathbf{A})$. Then the *finite-sample breakdown point under cellwise contamination* of a precision matrix estimator $\hat{\boldsymbol{\Theta}}$ is defined as

$$\varepsilon_n(\hat{\boldsymbol{\Theta}}, \mathbf{X}) = \min_{m=1,\ldots,n} \{\frac{m}{n} : \sup_{\mathbf{X}^m} D(\hat{\boldsymbol{\Theta}}(\mathbf{X}), \hat{\boldsymbol{\Theta}}(\mathbf{X}^m)) = \infty\}, \tag{14}$$

where $\mathbf{X}^m$ denotes a corrupted sample obtained from $\mathbf{X} \in \mathbb{R}^{n \times p}$ by replacing in each column at most m cells by arbitrary values. The following theorem was proven in Öllerer and Croux (2015).

Theorem 1 *The finite-sample breakdown point under cellwise contamination of the robust precision matrix estimator $\hat{\boldsymbol{\Theta}}_{\mathbf{S}}(\mathbf{X})$ fulfills*

$$\varepsilon_n(\hat{\boldsymbol{\Theta}}_{\mathbf{S}}, \mathbf{X}) \geq \varepsilon_n^+(\mathbf{S}, \mathbf{X}) \tag{15}$$

with $\mathbf{S}$ a positive semidefinite covariance estimator.

Here we used the *explosion* finite-sample breakdown point under cellwise contamination of a covariance matrix estimate $\mathbf{S}$, defined as

$$\varepsilon_n^+(\mathbf{S}, \mathbf{X}) = \min_{m=1,\ldots,n} \{\frac{m}{n} : \sup_{\mathbf{X}^m} |\lambda_1(\mathbf{S}(\mathbf{X})) - \lambda_1(\mathbf{S}(\mathbf{X}^m))| = \infty\}, \tag{16}$$

where $\mathbf{X}^m$ denotes a corrupted sample obtained from $\mathbf{X}$ by replacing in each column at most m cells by arbitrary values. Theorem 1 shows that Glasso preserves the robustness of the initial estimator. Moreover, Glasso prevents by construction explosion of the precision matrix estimator, and one only needs explosion robustness of the input covariance matrix $\mathbf{S}$.

Consider now our proposal for $\mathbf{S}$, where

$$s_{jk} = \text{scale}(\mathbf{x}^j)\,\text{scale}(\mathbf{x}^k) r(\mathbf{x}^j, \mathbf{x}^k) \qquad j, k = 1, \ldots, p.$$

It was shown in Öllerer and Croux (2015) that the explosion breakdown point under cellwise contamination of $\mathbf{S}$ is always larger than the explosion breakdown point of the scale estimator used. The Q_n-estimator has an explosion breakdown point of 50 %, resulting in a breakdown point of 50 % under cellwise contamination for the two-step estimators of Sect. 2. But the correlation measure r in the above definition may be the transformed Quadrant, Kendall, or Spearman correlation given in (9), (10) and (11). In these cases, the three-step estimator discussed in Sect. 2.2 needs to be used. The following result generalizes Proposition 1 in Öllerer and Croux (2015).

Proposition 1 *Let* $\mathbf{S}$ *be the covariance estimator based on pairwise correlations as defined in (4). Then*

$$\varepsilon_n^+(\mathbf{S}_{perturb}, \mathbf{X}) \geq \max_{j=1,\ldots,p} \varepsilon_n^+(\text{scale}, \mathbf{x}^j) \quad \textit{and} \quad \varepsilon_n^+(\mathbf{S}_{npd}, \mathbf{X}) \geq \max_{j=1,\ldots,p} \varepsilon_n^+(\text{scale}, \mathbf{x}^j), \tag{17}$$

with $\varepsilon_n^+(\text{scale}, \mathbf{x}^j)$ *the explosion breakdown point of the scale estimator used.*

Proof We first proof the result for the perturbation method. Using the triangular inequality, we obtain

$$|\lambda_1(\mathbf{S}_{perturb}(\mathbf{X})) - \lambda_1(\mathbf{S}_{perturb}(\mathbf{X}^m))| \leq |\lambda_1(\mathbf{S}_{perturb}(\mathbf{X}))| + |\lambda_1(\mathbf{S}_{perturb}(\mathbf{X}^m))|. \tag{18}$$

From Definition (12) we get

$$\lambda_1(\mathbf{S}_{perturb}(\mathbf{X}^m)) = \lambda_1(\mathbf{S}(\mathbf{X}^m)) - \min(0, \lambda_p(\mathbf{S}(\mathbf{X}^m))). \tag{19}$$

Using a result from Algebra (see Seber 2008, Eq. 6.26a), we have

$$|\lambda_r(\mathbf{S}(\mathbf{X}^m))| \leq p \max_{i,j=1,\ldots,p} |\mathbf{S}(\mathbf{X^m})_{ij}| \leq p \max_{j,k=1,\ldots,p} \text{scale}((\mathbf{X}^m)^j)\,\text{scale}((\mathbf{X}^m)^k) \tag{20}$$

for all $r = 1, \ldots, p$ and any $m \in \{1, \ldots, n\}$, where $(\mathbf{X}^m)^j$ denotes the jth column of matrix $\mathbf{X}^m$. For the second inequality in (20) we use the fact that the correlation measures (also the transformed ones) have an absolute value smaller than 1.

Equations (18), (19) and (20), together with the definition of the explosion breakdown point (16) show that (17) holds. The proof for the npd method is analogous, and even more simple. Indeed, it follows immediately from (13) that

$$\lambda_1(\mathbf{S}_{npd}(\mathbf{X}^m)) = \lambda_1(\mathbf{S}(\mathbf{X}^m)),$$

where we note that a matrix with nonnegative values on the diagonal must have a nonnegative largest eigenvalue.

The proposition above combined with Theorem 1 shows that also the three-stage estimators have an explosion breakdown point under cellwise contamination of at least 50 %.

5 Simulations

In this section, we perform a simulation study to compare the performance of the different precision matrix estimators introduced in Sect. 2. We compare the consistent three-step estimators of Sect. 2.2 to the inconsistent two-step estimator of Sect. 2.1. For the former, we use both methods for making the symmetric covariance matrix positive semidefinite: the nearest positive definite matrix (npd) method and the perturbation method. We also include the consistent two-step estimator based on the Gaussian rank correlation (7), for which no third step is needed. As a benchmark, we compare with the nonrobust estimators Glasso, where the sample covariance matrix is taken as an input in (2), and with the inverse of the sample covariance matrix (that can only be computed if $n > p$).

For the sake of comparison, we also add the spatial sign correlation matrix of Dürre et al. (2015). It is based on pairwise spatial correlations, which are then transformed such that the final estimator is consistent at the normal distribution. As a result, the spatial sign correlation matrix is not necessarily positive semidefinite, and either the npd method or the perturbation method need to be applied prior to using the matrix in Glasso, thus, it is a three-step procedure. The spatial sign correlation matrix can be computed with the function `sscor` of the R-package `sscor`. One could also look at the two-step procedure, the spatial sign correlations without the transformation, but since this is not implemented in R directly, we limit ourselves to the three-step approach.

The setup of the simulation study is taken over from Öllerer and Croux (2015). We use four sampling schemes to cover different patterns of the precision matrix $\boldsymbol{\Theta}_0 \in \mathbb{R}^{p \times p}$

- 'banded': $(\boldsymbol{\Theta}_0)_{ij} = 0.6^{|i-j|}$
- 'sparse': $\boldsymbol{\Theta}_0 = \mathbf{B} + \delta \mathbf{I}_p$ with $\mathbb{P}[b_{ij} = 0.5] = 0.1$ and $\mathbb{P}[b_{ij} = 0] = 0.9$ for $i \neq j$. The parameter δ is chosen such that the condition number of $\boldsymbol{\Theta}_0$ equals p. Then the matrix is standardized to have unit diagonals. The same matrix is used for all simulation runs.
  To find the specific value of δ, we numerically solve the equation $\kappa(\boldsymbol{\Theta}_0) = p$, where κ denotes the condition number. Using a smaller condition number would create a matrix more similar to the identity matrix, using a larger condition number would run the risk of not having a positive definite matrix. Specifically, the nonzero elements of $\boldsymbol{\Theta}_0$ are 0.246 for $p = 30$ and 0.150 for $p = 100$.
- 'dense': $(\boldsymbol{\Theta}_0)_{ii} = 1$ and $(\boldsymbol{\Theta}_0)_{ij} = 0.5$ for $i \neq j$
- 'diagonal': $(\boldsymbol{\Theta}_0)_{ii} = 1$ and $(\boldsymbol{\Theta}_0)_{ij} = 0$ for $i \neq j$

For each sampling scheme, we generate $M = 1000$ samples of size $n = 50$ from a multivariate normal $\mathcal{N}(0, \boldsymbol{\Theta}_0^{-1})$. We take as dimensions $p = 3, p = 30$ and $p = 100$. To each data set, we then add 0, 5 and 10 % of cellwise contamination. This means that we randomly select 0, 5 and 10 % of the cells and replace their value with a draw from a normal distribution $\mathcal{N}(10, 0.2)$.

Table 2 Simulation results: Kullback–Leibler criterion (KL) for the banded simulation setup with $n = 50$ averaged over $M = 1000$ simulations using the BIC criterion to select λ

	$p = 3$			$p = 30$			$p = 100$		
% outliers	0	5	10	0	5	10	0	5	10
2-step Quadrant	0.59	0.82	1.07	11.08	13.29	15.78	38.71	46.83	55.49
3-step Quadrant (npd)	0.34	0.55	0.81	12.42	15.11	17.97	49.50	60.62	70.92
3-step Quadrant (perturb)	0.34	0.55	0.81	16.01	19.37	22.69	63.56	77.41	90.12
2-step Spearman	0.30	0.63	0.98	10.71	13.38	15.87	38.58	47.11	55.64
3-step Spearman (npd)	0.28	0.60	0.95	10.66	13.47	15.99	39.44	48.27	56.98
3-step Spearman (perturb)	0.28	0.60	0.95	10.78	13.50	16.00	40.67	49.78	58.77
2-step Kendall	0.56	0.91	1.19	10.80	12.70	14.80	36.73	43.65	51.12
3-step Kendall (npd)	0.27	0.58	0.93	10.71	13.52	16.08	40.09	49.00	57.84
3-step Kendall (perturb)	0.27	0.58	0.93	11.25	13.92	16.39	46.34	56.46	66.31
3-step Spatial Sign (npd)	0.31	0.54	0.85	11.99	14.41	16.97	47.10	55.96	65.26
3-step Spatial Sign (perturb)	0.31	0.54	0.85	15.76	18.37	21.05	70.69	82.77	95.00
2-step Gaussian Rank	0.27	0.68	1.04	10.63	13.40	15.78	38.59	47.09	55.42
Glasso	0.23	2.98	4.11	10.32	30.55	42.48	38.01	106.53	145.67
Sample Covariance	0.14	2.40	3.54	39.39	27.06	31.11			

We compare the performance of the different estimators $\hat{\boldsymbol{\Theta}}$ by the Kullback–Leibler (KL) divergence (see e.g., Bühlmann and van de Geer 2011)

$$KL(\hat{\boldsymbol{\Theta}}, \boldsymbol{\Theta}_0) = \operatorname{tr}(\boldsymbol{\Theta}_0^{-1}\hat{\boldsymbol{\Theta}}) - \log\det(\boldsymbol{\Theta}_0^{-1}\hat{\boldsymbol{\Theta}}) - p.$$

The lower the value of KL, the better for the estimate. The results for the banded simulation setup are given in Table 2. The standard errors around all reported results are smaller than 3 % of the reported numbers. Let us focus on what is new in this simulation study compared to Öllerer and Croux (2015).

(i) For $p = 3$, the inconsistent two-step Quadrant estimator results in a substantially higher KL-value than the consistent three-step Quadrant estimators. Here, the additional step leads to a considerable improvement of the estimate. However, for higher values of p, the inconsistent two-step Quadrant estimator yields lower values of KL. This is a surprising outcome: rendering the Quadrant correlation-based estimator consistent comes at the price of increased Kullback–Leibler

distance, at least for the configurations of interest in this paper (n close to or smaller than p). The same can be observed for the Kendall correlation.
For the Spearman estimator, there is not much difference between the two-step and three-step method (at least not when using npd). This was to be expected (see Fig. 1).

(ii) Looking at the Kendall-based estimators we observe the following: The consistent three-step Kendall estimators perform similarly to the consistent three-step Spearman estimators in terms of KL. The inconsistent two-step Kendall estimator leads in low dimensions ($p = 3$) to high values of KL compared to the other robust estimators. In high dimensions, however, the situation is different with slightly lower values of KL than the other estimators.

(iii) Comparing the perturbation method and the nearest positive definite (npd) approach, the npd has a clear advantage. Particularly for high dimensions the difference is pronounced (for the estimators based on Quadrant, Kendall, and spatial sign).

(iv) The spatial sign correlation matrix leads to higher values of KL than the other estimators, especially when p is large compared to n.

Comparing the different estimators in Table 2 results in the following findings:

(i) If no outliers are present, and if n is close to p, then Glasso based on the sample covariance matrix is best. But the difference to the two-step Kendall, Gaussian rank, and Spearman-based estimators is small. The quadrant correlation is much less efficient for clean normally distributed data.

(ii) Under contamination, the nonrobust Glasso and the sample covariance matrix are not reliable anymore, and have much higher values of KL. For $p = 30$ and $p = 100$, best results are achieved by the two-step Kendall estimator, closely followed by the two-step estimators based on Gaussian rank, Quadrant, and Spearman correlations. There do not seem to be major differences in performance between the latter methods in these configurations. This may partly be explained by the fact that the pairwise covariances computed in (4) use the same robust scale estimator.
In the low-dimensional setting ($p = 3$) under contamination, we have the following relations for the consistent estimators: Quadrant is better than Kendall, which is on its turn better than Spearman, which is better than Gaussian Rank.

Result of KL for the other three simulation setups are given in Table 3. Again, the standard errors around the reported results are smaller than 3 % of the reported numbers. For the "dense" setting exactly the same conclusions can be drawn as for the "banded" setting of Table 2. For the other two settings, which are characterized by a sparse true precision matrix, we see that Glasso outperforms the sample covariance matrix even for $p = 3$. The overall conclusion of these simulation results is that the two-step Gaussian rank, the two-step Kendall, the two-step Spearman, the two-step Quadrant, and the three-step Spearman (npd) are comparable and yield the best results.

Table 3 Same as Table 2, for the three other simulation setups

	% outliers	$p = 3$			$p = 30$			$p = 100$		
		0	5	10	0	5	10	0	5	10
Sparse	2-step Quadrant	0.11	0.25	0.42	7.83	10.12	12.59	36.39	44.95	53.47
	3-step Quadrant (npd)	0.15	0.28	0.44	9.16	12.00	14.76	48.02	59.25	69.73
	3-step Quadrant (perturb)	0.15	0.28	0.44	13.02	16.43	19.75	63.68	77.94	90.41
	2-step Spearman	0.11	0.25	0.42	7.97	10.38	12.81	37.53	45.79	54.05
	3-step Spearman (npd)	0.11	0.25	0.43	8.01	10.48	12.95	38.51	47.00	55.44
	3-step Spearman (perturb)	0.11	0.25	0.43	8.03	10.50	12.96	39.84	48.67	57.47
	2-step Kendall	0.09	0.22	0.38	7.54	9.55	11.67	34.45	41.60	48.98
	3-step Kendall (npd)	0.11	0.25	0.43	8.04	10.51	13.02	39.19	47.83	56.33
	3-step Kendall (perturb)	0.11	0.25	0.43	8.33	10.76	13.28	45.34	55.48	65.10
	3-step Spatial Sign (npd)	0.14	0.27	0.44	8.83	11.34	13.89	45.59	55.18	64.03
	3-step Spatial Sign (perturb)	0.13	0.27	0.44	12.69	15.41	18.06	69.89	82.88	94.48
	2-step Gaussian Rank	0.11	0.25	0.42	8.04	10.37	12.75	37.64	45.58	53.77
	Glasso	0.09	3.15	4.58	7.94	34.42	48.30	37.21	120.76	166.08
	Sample Covariance	0.14	2.70	4.04	39.39	27.10	33.77			
Dense	2-step Quadrant	0.56	0.76	0.94	4.38	6.53	8.92	11.35	19.53	28.08
	3-step Quadrant (npd)	0.34	0.51	0.74	5.68	8.29	11.07	21.84	32.84	43.41
	3-step Quadrant (perturb)	0.34	0.51	0.74	8.97	12.29	15.57	35.15	49.81	62.43
	2-step Spearman	0.29	0.68	0.94	4.40	6.53	8.94	11.40	19.47	28.02
	3-step Spearman (npd)	0.26	0.64	0.92	4.47	6.65	9.07	12.13	20.48	29.27
	3-step Spearman (perturb)	0.26	0.64	0.92	4.56	6.66	9.08	13.17	21.94	31.06
	2-step Kendall	0.64	0.81	0.97	3.83	5.66	7.82	8.88	15.75	23.39
	3-step Kendall (npd)	0.25	0.62	0.91	4.50	6.67	9.13	12.63	21.16	30.06
	3-step Kendall (perturb)	0.25	0.62	0.91	4.77	6.84	9.30	17.76	28.19	38.20

(continued)

Table 3 (continued)

	% outliers	$p = 3$			$p = 30$			$p = 100$		
		0	5	10	0	5	10	0	5	10
Dense	3-step Spatial Sign (npd)	0.32	0.55	0.84	5.25	7.53	9.87	19.08	28.01	37.27
	3-step Spatial Sign (perturb)	0.32	0.55	0.84	8.36	10.82	13.40	41.68	54.57	66.95
	2-step Gaussian Rank	0.26	0.70	0.95	4.39	6.51	8.90	11.38	19.41	27.96
	Glasso	0.20	2.97	4.18	4.22	24.97	37.29	10.72	82.75	123.32
	Sample Covariance	0.14	2.44	3.63	39.39	23.54	26.43			
Diagonal	2-step Quadrant	0.11	0.25	0.42	1.92	4.08	6.51	7.82	15.89	24.60
	3-step Quadrant (npd)	0.15	0.28	0.44	3.22	5.82	8.62	18.21	29.33	40.01
	3-step Quadrant (perturb)	0.15	0.28	0.44	6.37	9.78	13.16	31.74	46.17	58.78
	2-step Spearman	0.11	0.25	0.42	1.91	4.07	6.49	7.73	15.78	24.41
	3-step Spearman (npd)	0.11	0.25	0.43	2.00	4.19	6.65	8.50	16.88	25.71
	3-step Spearman (perturb)	0.11	0.25	0.43	2.00	4.19	6.66	9.46	18.38	27.49
	2-step Kendall	0.09	0.22	0.38	1.36	3.18	5.38	5.24	12.15	19.81
	3-step Kendall (npd)	0.11	0.25	0.43	2.03	4.22	6.68	8.98	17.55	26.54
	3-step Kendall (perturb)	0.11	0.25	0.43	2.11	4.33	6.83	14.22	24.53	34.62
	3-step Spatial Sign (npd)	0.14	0.27	0.44	2.79	5.03	7.53	15.34	24.40	33.72
	3-step Spatial Sign (perturb)	0.13	0.27	0.44	5.70	8.32	11.05	38.03	51.01	63.11
	2-step Gaussian Rank	0.11	0.25	0.42	1.91	4.06	6.47	7.71	15.79	24.38
	Glasso	0.09	3.15	4.58	1.74	35.05	50.27	7.00	120.26	173.86
	Sample Covariance	0.14	2.70	4.04	39.39	26.94	34.38			

6 Graphical Models

Sparse estimation of the precision matrix has a direct application in graphical modeling. If element (i,j) of $\hat{\boldsymbol{\Theta}}$ equals zero, then the estimated partial correlation between variables i and j equals zero. Since we are assuming normality, this means that variables i and j are independent, conditional on the other variables. The variables are represented by the nodes of the graph, and if two variables are estimated as conditionally dependent, an undirected arrow is drawn between the corresponding nodes. The rank-based correlation coefficient matrices, Spearman, Kendall, or Gaussian rank, can then be used as an input for the Glasso method. Several papers discussed this approach in depth, see Liu et al. (2009), Liu et al. (2012a), Xue and Zou (2012), Zhao et al. (2014b). They point out an important advantage of using rank-based correlation. If the distribution is only multivariate normal after monotone transformation of the variables (then the distribution is said to be "nonparanormal," and it has a multivariate Gaussian copula), zero partial correlation still implies conditional independence.

Other robust approaches to graphical modeling have been considered: Kalisch and Bühlmann (2008) robustly estimate the partial correlation of different subsets of variables and then test if they are zero. Finegold and Drton (2011) show how a conditional independency graph can be estimated for the more robust family of t-distributions. Vogel and Fried (2011) formulate a suitable graphical model for the whole elliptical family, where graphs model partial correlations (and not anymore conditional independencies). Another approach for complex elliptical symmetric distributions has been developed by Ollila and Tyler (2014). They introduce a regularized M-estimator of scatter that is unique and consistent. Elliptical graphical modeling has been extended to the so called *transelliptical* or *meta-elliptical* family by Liu et al. (2012b) and Bilodeau (2014). A major difference with all of these papers is that we study robust and sparse inverse covariance matrices, and do not confine ourselves to correlation matrices. Obviously, in the context of graphical modeling, the retrieved graph will be exactly the same.

To measure how well the graph structure is recovered, we compute false positive (FP) and false negative (FN) rates

$$\text{FP} = \frac{|\{(i,j) : i = 1, \ldots, n; j = 1, \ldots, p : (\hat{\boldsymbol{\Theta}})_{ij} \neq 0 \wedge (\boldsymbol{\Theta}_0)_{ij} = 0\}|}{|\{(i,j) : i = 1, \ldots, n; j = 1, \ldots, p : (\boldsymbol{\Theta}_0)_{ij} = 0\}|}$$

$$\text{FN} = \frac{|\{(i,j) : i = 1, \ldots, n; j = 1, \ldots, p : (\hat{\boldsymbol{\Theta}})_{ij} = 0 \wedge (\boldsymbol{\Theta}_0)_{ij} \neq 0\}|}{|\{(i,j) : i = 1, \ldots, n; j = 1, \ldots, p : (\boldsymbol{\Theta}_0)_{ij} \neq 0\}|}$$

They give the percentage of zero-elements of the precision matrix wrongly estimated as nonzero and the percentage of nonzero elements that are wrongly estimated as zero. In other words, FN gives the percentage of undetected edges of the graph, and FP the percentage of falsely detected edges. The lower these values are, the better.

To investigate how well the different estimators are able to recover the graph structure, Table 4 gives FP and FN for the setups $p = 30$ and $p = 100$ in the "sparse" setting. The inverse sample covariance matrix is a nonsparse estimator, and therefore always leads to an FP equal to one and an FN equal to zero. The other estimators lead to pretty similar values of FP and FN. The nonrobust Glasso for $p = 30$ has an increased FN rate under contamination. The three-step Spearman and Kendall yield the lowest FP and FN rates, but differences to the other procedures are small.

Note that Table 4 presents a relative comparison of the different methods and not a qualitative evaluation of the false negative rate. The reason is that BIC, even though it is fast, usually selects too sparse graphs. Using other selection procedures such as cross validation (see Öllerer and Croux 2015) can improve results, but is computationally more expensive. Research on good high-dimensional selection procedures for λ is still ongoing (see e.g. Abbruzzo et al. 2014; Liu et al. 2010; Foygel and Drton 2010). Therefore, we stick for simplicity to the well established BIC criterion.

7 Discussion

We discuss robust and sparse estimators of the precision matrix, computable in high dimensions with $p > n$. This proceedings paper complements Öllerer and Croux (2015), but we provide further discussion and study additionally the consistent versions of estimators based on Quadrant correlation, Spearman's rank correlation, and Kendall correlation. For computing the latter estimators, an additional step is needed to guarantee positive definiteness of the matrices. We prove that this extra step is not distorting the high breakdown point of the estimators.

The estimators discussed in this paper are using sign and rank correlation measures. Spearman and Kendall correlation provide a good trade-off between robustness and efficiency. In Croux and Dehon (2010) it was shown that Spearman and Kendall correlation behave rather similarly in the bivariate setting.

As shown in Liu et al. (2009), the estimators proposed in this paper consistently recover the underlying graph in a nonparanormal model, which includes the normal model. Since the consistency transformation of the Kendall and Quadrant correlations still hold in the broader classes of transelliptical or metaellipitcal distributions (Liu et al. 2012b; Bilodeau 2014), the methods based on the transformed Kendall and Quadrant correlations consistently recover the underlying graph structure also in these models. Furthermore, also the correlation matrix will be estimated consistently. However, when transforming the correlation matrix to a covariance matrix in Eq. (4), the consistency of the scale estimator also needs to be taken into account. Since the scale estimator Q_n is consistent only at the normal model, consistency for the covariance matrix and its inverse requires multivariate normality.

While we focused our attention on the estimation of the precision matrix and the covariance matrix, we did not consider the estimation of the location parameter $\boldsymbol{\mu}$ yet. Note that estimation based on Spearman correlation does not require an auxiliary location estimate. A simple robust estimator for $\boldsymbol{\mu}$ is the coordinatewise median,

Table 4 Simulation results: False Positive Rate (FP) and False Negative Rate (FN) for the sparse simulation setup with $n = 50$ averaged over $M = 1000$ simulations using BIC criterion to select λ

	$p = 30$						$p = 100$					
% outliers	0		5		10		0		5		10	
	FP	FN	FP	FN	FP	FN	FP	FN	FP	FN	FP	FN
2-step Quadrant	0.01	0.70	0.01	0.71	0.01	0.72	0.00	0.90	0.00	0.90	0.00	0.90
3-step Quadrant (npd)	0.01	0.67	0.01	0.68	0.01	0.70	0.00	0.90	0.00	0.90	0.00	0.90
3-step Quadrant (perturb)	0.01	0.70	0.01	0.71	0.01	0.72	0.00	0.90	0.00	0.90	0.00	0.91
2-step Spearman	0.01	0.64	0.01	0.68	0.01	0.71	0.00	0.89	0.00	0.90	0.00	0.90
3-step Spearman (npd)	0.01	0.63	0.01	0.68	0.01	0.71	0.00	0.89	0.00	0.90	0.00	0.90
3-step Spearman (perturb)	0.01	0.63	0.01	0.68	0.01	0.71	0.00	0.89	0.00	0.90	0.00	0.90
2-step Kendall	0.00	0.70	0.00	0.72	0.00	0.73	0.00	0.90	0.00	0.90	0.00	0.91
3-step Kendall (npd)	0.01	0.63	0.01	0.67	0.01	0.70	0.00	0.89	0.00	0.90	0.00	0.90
3-step Kendall (perturb)	0.01	0.64	0.01	0.68	0.01	0.71	0.00	0.89	0.00	0.90	0.00	0.90
3-step Spatial Sign (npd)	0.01	0.67	0.01	0.68	0.01	0.71	0.00	0.90	0.00	0.90	0.00	0.90
3-step Spatial Sign (perturb)	0.00	0.70	0.00	0.71	0.01	0.73	0.00	0.91	0.00	0.91	0.00	0.91
2-step Gaussian Rank	0.01	0.64	0.01	0.69	0.01	0.72	0.00	0.89	0.00	0.90	0.00	0.90
Glasso	0.00	0.64	0.01	0.76	0.01	0.76	0.00	0.89	0.01	0.91	0.00	0.91
Sample Covariance	1.00	0.00	1.00	0.00	1.00	0.00						

which simply computes the median for every variable separately. Obviously, this estimator is highly robust and computable in high dimensions. However, this estimator is not affine equivariant, and neither are the covariance matrix estimators $\mathbf{S}$ considered in this paper. If we transform the observation $\mathbf{x}_i$ into $\mathbf{A}\mathbf{x}_i + \mathbf{b}$, with $\mathbf{A}$ a non-singular matrix and $\mathbf{b}$ a constant vector, then the estimators $\hat{\boldsymbol{\mu}}$ and $\hat{\boldsymbol{\Sigma}}$ are said to be affine equivariant if they change accordingly to $\mathbf{A}\hat{\boldsymbol{\mu}} + b$ and $\mathbf{A}\hat{\boldsymbol{\Sigma}}\mathbf{A}^\top$. We only have this property for diagonal matrices $\mathbf{A}$.

A popular robust estimator of location and covariance is the Minimum Covariance Determinant (MCD) estimator (Rousseeuw and Van Driessen 1999) which is affine equivariant, but ill defined if $p > n$. Indeed, the MCD is looking for a subsample of half the sample size having smallest value of the determinant of the covariance matrix computed from this subsample. But if $p > n$, or even if $p > n/2$ all the determinants of covariance matrices computed from halfsamples are zero, and it is not clear what to do then. Moreover, the MCD estimator is not robust to cellwise outliers if you have many of them, as is common in high dimension. There is recent work of Agostinelli et al. (2015) proposing an almost affine equivariant, location/covariance matrix estimator robust to cellwise contamination. Unfortunately, the latter estimator is not computable if $p > n$, as is the proposal of Van Aelst et al. (2011). To sum up, one needs to give up affine equivariance when constructing robust estimators for $p > n$. We refer to Alqallaf et al. (2009) and Tyler (2010) for further discussion on equivariance properties and contamination models appropriate in high dimensions.

Robust correlation matrices based on pairwise rank correlation estimators have been studied before in the literature. In Sect. 6 we reviewed their use in graphical modeling. In principal component analysis they have been used by Van Aelst et al. (2010), who used Spearman correlation. Alqallaf et al. (2002) use Quadrant correlation for non-sparse covariance matrix estimation. We believe that the cellwise robust covariance matrix estimators based on ranks and signs and discussed in this and other papers have a lot of potential for high-dimensional data analysis.

Acknowledgments The authors wish to acknowledge the support from the GOA/12/014 project of the Research Fund KU Leuven. We also would like to thank the referees for their constructive comments that improved the paper considerably.

References

Abbruzzo A, Vujacic I, Wit E, Mineo A (2014) Generalized information criterion for model selection in penalized graphical models. arXiv:1403.1249

Agostinelli C, Leung A, Yohai V, Zamar R (2015) Robust estimation of multivariate location and scatter in the presence of cellwise and casewise contamination. Test 24(3):441–461

Alqallaf F, Konis K, Martin R, Zamar R (2002) Scalable robust covariance and correlation estimates for data mining. In: Proceedings of the eighth ACM SIGKDD international conference on knowledge discovery and data mining, ACM, pp 14–23

Alqallaf F, Van Aelst S, Yohai V, Zamar R (2009) Propagation of outliers in multivariate data. Ann Stat 37(1):311–331

Bilodeau M (2014) Graphical lassos for meta-elliptical distributions. Can J Stat 42:185–203

Boudt K, Cornelissen J, Croux C (2012) The Gaussian rank correlation estimator: robustness properties. Stat Comput 22(2):471–483

Bühlmann P, van de Geer S (2011) Statistics for high-dimensional data. Springer, Heidelberg

Croux C, Dehon C (2010) Influence functions of the Spearman and Kendall correlation measures. Stat Meth Appl 19(4):497–515

Dürre A, Vogel D, Fried R (2015) Spatial sign correlation. J Multivar Anal 135:89–105

Finegold M, Drton M (2011) Robust graphical modeling of gene networks using classical and alternative t-distributions. Ann Appl Stat 5(2A):1057–1080

Foygel R, Drton M (2010) Extended bayesian information criteria for gaussian graphical models. In: Advances in neural information processing systems 23, Curran Associates, Inc., pp 604–612

Friedman J, Hastie T, Tibshirani R (2008) Sparse inverse covariance estimation with the graphical lasso. Biostatistics 9(3):432–441

Gnanadesikan R, Kettenring J (1972) Robust estimates, residuals and outlier detection with multiresponse data. Biometrics 28(1):81–124

Higham N (2002) Computing the nearest correlation matrix - a problem from finance. IMA J Numer Anal 22(3):329–343

Kalisch M, Bühlmann P (2008) Robustification of the pc-algorithm for directed acyclic graphs. J Comput Graph Stat 17(4):773–789

Kendall M (1938) A new measure of rank correlation. Biometrika 30:81–93

Liu H, Lafferty J, Wasserman L (2009) The nonparanormal: semiparametric estimation on high dimensional undirected graphs. J Mach Learn Res 10:2295–2328

Liu H, Roeder K, Wasserman L (2010) Stability approach to regularization selection (StARS) for high dimensional graphical models. In: Advances in neural information processing systems 23, Curran Associates, Inc., pp 1432–1440

Liu H, Han F, Yuan M, Lafferty J, Wasserman L (2012a) High-dimensional semiparametric Gaussian copula graphical models. Ann Stat 40(4):2293–2326

Liu H, Han F, Zhang C (2012b) Transelliptical graphical models. In: Advances in neural information processing systems 25, Curran Associates, Inc., pp 800–808

Maronna R, Martin R, Yohai V (2006) Robust statistics, 2nd edn. Wiley, Hoboken

Öllerer V, Croux C (2015) Robust high-dimensional precision matrix estimation. In: Nordhausen K, Taskinen S (eds) Modern Nonparametric, Robust and Multivariate Methods, Springer, pp 325–350

Ollila E, Tyler D (2014) Regularized M-estimators of scatter matrix. IEEE Trans Signal Process 62(22):6059–6070

Rousseeuw P, Croux C (1993) Alternatives to the median absolute deviation. J Am Stat Assoc 88(424):1273–1283

Rousseeuw P, Molenberghs G (1993) Transformation of nonpositive semidefinite correlation matrices. Commun Stat - Theory Meth 22(4):965–984

Rousseeuw P, Van Driessen K (1999) A fast algorithm for the minimum covariance determinant estimator. Technometrics 41(3):212–223

Rousseeuw P, Croux C, Todorov V, Ruckstuhl A, Salibian-Barrera M, Verbeke T, Koller M, Maechler M (2015) Robustbase: basic robust statistics. http://CRAN.R-project.org/package=robustbase, r package version 0.92-3

Seber G (2008) A matrix handbook for Statisticians. Wiley, Hoboken

Tarr G, Müller S, Weber N (2016) Robust estimation of precision matrices under cellwise contamination. Comput Stat Data Anal 93:404–420

Todorov V, Filzmoser P, Fritz H, Kalcher K (2014) pcaPP: Robust PCA by Projection Pursuit. http://CRAN.R-project.org/package=pcaPP, r package version 1.9-60

Tyler D (2010) A note on multivariate location and scatter statistics for sparse data. Stat Probab Lett 80(17–18):1409–1413

Van Aelst S, Vandervieren E, Willems G (2010) Robust principal component analysis based on pairwise correlation estimators. In: Proceedings of COMPSTAT2010, Physica-Verlag HD, pp 573–580

Van Aelst S, Vandervieren E, Willems G (2011) Stahel-Donoho estimators with cellwise weights. J Stat Comput Simul 81(1):1–27

Vogel D, Fried R (2011) Elliptical graphical modelling. Biometrika 98(4):935–951

Xue L, Zou H (2012) Regularized rank-based estimation of high-dimensional nonparanormal graphical models. Ann Stat 40(5):2541–2571

Yuan M, Lin Y (2007) Model selection and estimation in the Gaussian graphical model. Biometrika 94(1):19–35

Zhao T, Liu H, Roeder K, Lafferty J, Wasserman L (2012) The `huge` package for high-dimensional undirected graph estimation in R. J Mach Learn Res 13:1059–1062

Zhao T, Liu H, Roeder K, Lafferty J, Wasserman L (2014a) huge: High-dimensional undirected graph estimation. URL http://CRAN.R-project.org/package=huge, r package version 1.2.6

Zhao T, Roeder K, Liu H (2014b) Positive semidefinite rank-based correlation matrix estimation with application to semiparametric graph estimation. J Comput Graph Stat 23(4):895–922

Robust Mixture Regression Using Mixture of Different Distributions

Fatma Zehra Doğru and Olcay Arslan

1 Introduction

Mixture regression models were first introduced by Quandt (1972), Quandt and Ramsey (1978) as switching regression models, which are used to explore the relationship between variables that come from some unknown latent groups. These models are widely applied in areas such as engineering, genetics, biology, econometrics and marketing. These mixture regression models are used to model data sets which contain heterogeneous groups. Figure 1 shows the scatter plots of this type of real data sets used in literature. A pure fundamental tone electronically obtained overtones added was played to a trained musician in the tone perception data which given by Cohen (1984) in Fig. 1a. The overtones were determined by a stretching ratio which is between the adjusted tone and the fundamental tone. 150 trials were performed by the same musicians in this experiment. This experiment was to reveal how the tuning ratio affects the perception of the tone and to choose if either of two musical perception theories was reasonable (see Cohen 1984 for more detail). The other data contains a number of green peach aphids which were released at various times over 51 small tobacco plants (used as surrogates for potato plants) and the number of infected plants was recorded after each release given in Fig. 1b (see Turner 2000 for more detailed explanations). From these figures, we can observe that there are two groups in both examples. Therefore, these data sets should be modeled by using the mixture regression models.

In general, the parameters of a mixture regression model are estimated under normality assumption. Since the estimators based on normal distribution are sensitive to the outliers, robust mixture regression models have been proposed by Bai (2010) and Bai et al. (2012) to estimate the parameters of mixture regression using the M-estimation method. Wei (2012), Yao et al. (2014) proposed the mixture regression

---

F.Z. Doğru (✉)
Faculty of Arts and Sciences, Department of Statistics, Giresun University,
28100 Giresun, Turkey
e-mail: fatma.dogru@giresun.edu.tr

O. Arslan
Faculty of Science, Department of Statistics, Ankara University, 06100 Ankara, Turkey
e-mail: oarslan@ankara.edu.tr

C. Agostinelli et al. (eds.), *Recent Advances in Robust Statistics: Theory and Applications*, DOI 10.1007/978-81-322-3643-6_4

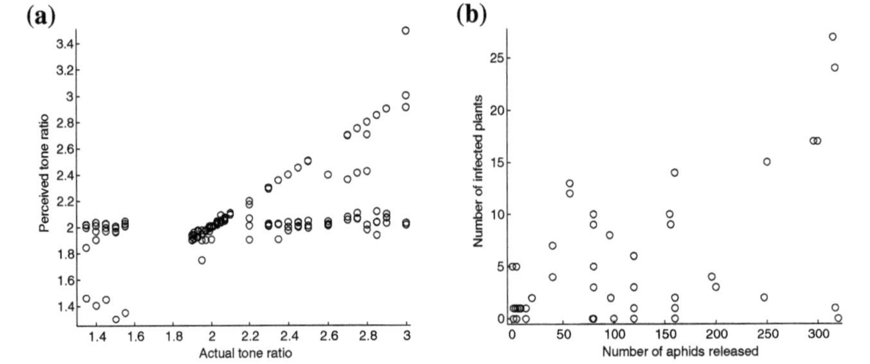

Fig. 1 **a** The scatter plot of the tone perception data. **b** The scatter plot of the aphids data

model based on the mixture of t distribution. Liu and Lin (2014) studied the mixture regression model based on the skew normal (Azzalini 1985, 1986) distribution. Doğru (2015), Doğru and Arslan (2016) propose a robust mixture regression procedure using the mixture of skew t distribution (Azzalini and Capitaino 2003) to model skewness and heavy-tailedness in the data with the groups.

Up to now mixture regression models are considered using the finite mixture of the same type of distributions such as mixture of normal or mixture of t distributions. The purpose of this work is to deal with the mixture regression model using the mixture of different type of distributions. This is due to the fact that the subclasses of data may not have same type of behavior. For example some of them may be heavy-tailed, skew or heavy-tailed skew. Using the same type of distributions to model such heterogeneous data may not produce efficient estimators. To accurately model this type of data we may need a mixture of distributions with different type of components. For example, it is clear that in the tone perception data (Fig. 1) two groups should have different type of error distributions. This is due to the fact that the observations around each line has differently scattered.

The rest of the paper is organized as follows. In Sect. 2, we give the mixture regression estimation based on mixture of different distributions. We consider two different mixtures. First, we consider the mixture of symmetric distributions. In particular, we take the mixture of normal and t distribution to estimate the regression parameters in a mixture regression model. Second model will be the mixture of skew distributions. In this context, we study the mixture of skew t and skew normal distribution to estimate the parameters of the mixture regression model. In both cases we give the EM algorithms in details. In Sect. 3, we provide a simulation study to demonstrate the performances of the proposed mixture regression estimators over the counterparts. In Sect. 4, we explore two real data examples to see the capability of the proposed estimators for real data sets. The paper is finalized with a conclusion section.

2 Mixture Regression Model Using the Mixture of Different Type of Distributions

In this section, we will carry out the mixture regression procedure based on the mixture of different distributions. We will only consider the mixture of two distributions, but mixture of more than two different types of distributions can be easily done using the methodology given in this paper.

2.1 *Mixture Regression Estimation Based on the Mixture of Normal and t Distributions*

A two-component mixture regression model can be defined as follows. Let Z be a latent class variable which is independent of explanatory variable $\mathbf{x}$. Then given $Z = i$, the response variable y and the p-dimensional explanatory variable $\mathbf{x}$ have the following linear model

$$y_j = \mathbf{x}_j'\boldsymbol{\beta}_i + \epsilon_i, i = 1, 2, \tag{1}$$

where $\mathbf{x}_j$ contains both the predictors and constant 1. Let $w_i = P(Z = i|\mathbf{x}), i = 1, 2$, be the mixing probability with $\sum_{i=1}^{2} w_i = 1$. The conditional density of y given $\mathbf{x}$ has the following form

$$f(y_j; \mathbf{x}_j, \boldsymbol{\Theta}) = w\phi\left(y_j; \mathbf{x}_j'\boldsymbol{\beta}_1, \sigma_1^2\right) + (1-w)f_t\left(y_j; \mathbf{x}_j'\boldsymbol{\beta}_2, \sigma_2^2, \nu\right), \tag{2}$$

where Z is not observed. This implies that the distribution of the first error term is a normal distribution with 0 mean and the variance σ_1^2 and the distribution of the second error term is a t distribution with 0 mean, the scale parameter σ_2^2 and the degrees of freedom ν. Let $\boldsymbol{\Theta} = (w, \boldsymbol{\beta}_1, \sigma_1^2, \boldsymbol{\beta}_2, \sigma_2^2, \nu)'$ be the vector of all the unknown parameters in the model (2).

The ML estimator of the unknown parameter $\boldsymbol{\Theta}$ is obtained by maximizing the following log-likelihood function

$$\ell(\boldsymbol{\Theta}) = \sum_{j=1}^{n} log(w\phi(y_j; \mathbf{x}_j'\boldsymbol{\beta}_1, \sigma_1^2) + (1-w)f_t(y_j; \mathbf{x}_j'\boldsymbol{\beta}_2, \sigma_2^2, \nu)). \tag{3}$$

However, the maximizer of the log-likelihood function does not have an explicit solution. Therefore, the numerical methods should be used to obtain the estimators for the parameters of interest. Because of the mixture structure of the model the EM algorithm (Dempster et al. 1977) will be the convenient numerical method to obtain the estimators for the parameters.

Let z_j be the latent variable with

$$z_j = \begin{cases} 1, \; if \; j^{th} \; observation \; is \; from \; 1^{th} \; component \\ 0, \; otherwise, \end{cases} \tag{4}$$

for $j = 1, \ldots, n$. The joint density function of two-component mixture regression model is

$$f(y_j, z_j) = \left(w\phi(y_j; \mathbf{x}_j'\boldsymbol{\beta}_1, \sigma_1^2)\right)^{z_j} \left((1-w) f_t(y_j; \mathbf{x}_j'\boldsymbol{\beta}_2, \sigma_2^2, \nu)\right)^{(1-z_j)}. \tag{5}$$

To further simplify the steps of the EM algorithm, we will use the scale mixture representation of the t distribution. Let the random variable u has a gamma distribution with the parameters $(\nu/2, \nu/2)$. Then, the conditional distribution of ϵ_2 given u will be $N(0, \sigma^2/u)$. With the scale mixture representation of the t distribution this joint density can be further simplified as

$$f\left(y_j, u_j, z_j\right) = \left(w \frac{1}{\sqrt{2\pi\sigma_1^2}} e^{-\frac{\left(y_j - \mathbf{x}_j'\boldsymbol{\beta}_1\right)^2}{2\sigma_1^2}} \right)^{z_j} \left((1-w) \frac{(\nu/2)^{\nu/2} u_j^{\frac{\nu}{2}-1} e^{-\frac{\nu}{2}u_j}}{\Gamma\left(\frac{\nu}{2}\right)\sqrt{2\pi\sigma_2^2/u_j}} e^{-\frac{\left(y_j - \mathbf{x}_j'\boldsymbol{\beta}_2\right)^2}{2\sigma_2^2/u_j}} \right)^{1-z_j}. \tag{6}$$

In this model, $(\mathbf{z}, \mathbf{u})$ are regarded as missing data and y is taken as observed data, where $\mathbf{y} = (y_1, \ldots, y_n)$, $\mathbf{u} = (u_1, \ldots, u_n)$ and $\mathbf{z} = (z_1, \ldots, z_n)$. Equation (6) is the joint density function of the complete data $(\mathbf{y}, \mathbf{u}, \mathbf{z})$. Using this joint density function the complete data log-likelihood function for $\boldsymbol{\Theta}$ can be written as follows

$$\begin{aligned} \ell(\boldsymbol{\Theta}; \mathbf{y}, \mathbf{u}, \mathbf{z}) = &\sum_{j=1}^{n} z_j \left(\log w - \frac{\log 2\pi}{2} - \frac{\log \sigma_1^2}{2} - \frac{\left(y_j - \mathbf{x}_j'\boldsymbol{\beta}_1\right)^2}{2\sigma_1^2} \right) \\ &+ (1 - z_j) \left(\log(1-w) - \frac{\log 2\pi}{2} - \frac{\log \sigma_2^2}{2} + \frac{\log u_j}{2} - \frac{\nu}{2} u_j \right. \\ &\left. - \frac{\left(y_j - \mathbf{x}_j'\boldsymbol{\beta}_2\right)^2}{2\sigma_2^2/u_j} - \log \Gamma\left(\frac{\nu}{2}\right) + \frac{\nu}{2} \log\left(\frac{\nu}{2}\right) + \left(\frac{\nu}{2} - 1\right) \log u_j \right). \end{aligned} \tag{7}$$

Since u_j and z_j for $j = 1, \ldots, n$, are taken as missing observations this log-likelihood function cannot be directly used to obtain the estimator for $\boldsymbol{\Theta}$. To overcome this latency problem we have to take the conditional expectation of the complete data log-likelihood function given y_j. This will be the E-step of the EM algorithm:
E-step:

$$\begin{aligned} E\left(\ell(\boldsymbol{\Theta}; \mathbf{y}, \mathbf{u}, \mathbf{z}) | y_j\right) = &\sum_{j=1}^{n} E\left(z_j | y_j\right) \left(\log w - \frac{\log 2\pi}{2} - \frac{\log \sigma_1^2}{2} - \frac{\left(y_j - \mathbf{x}_j'\boldsymbol{\beta}_1\right)^2}{2\sigma_1^2} \right) \\ &+ \left(1 - E\left(z_j | y_j\right)\right) \left(\log(1-w) - \frac{\log 2\pi}{2} - \frac{\log \sigma_2^2}{2} \right. \end{aligned}$$

$$+\frac{1}{2}E\left(\log u_j|y_j\right)-\frac{\left(y_j-\mathbf{x}_j'\boldsymbol{\beta}_2\right)^2E\left(u_j|y_j\right)}{2\sigma_2^2}-\frac{\nu}{2}E\left(u_j|y_j\right)$$
$$-\log\Gamma\left(\frac{\nu}{2}\right)+\frac{\nu}{2}\log\left(\frac{\nu}{2}\right)+\left(\frac{\nu}{2}-1\right)E\left(\log u_j|y_j\right)\Big). \tag{8}$$

To obtain this conditional expectation of the complete data log-likelihood function we have to find $\hat{z}_j = E(z_j|y_j, \hat{\boldsymbol{\Theta}})$, $\hat{u}_{1j} = E(u_j|y_j, \hat{\boldsymbol{\Theta}})$ and $\hat{u}_{2j} = E(\log u_j|y_j, \hat{\boldsymbol{\Theta}})$ given in (36), (37) and (38), where $\hat{\boldsymbol{\Theta}}$ is the current estimate for $\boldsymbol{\Theta}$.
The M-step of the EM algorithm will be as follows.

M-step: Maximize the following function with respect to $\boldsymbol{\Theta}$

$$Q\left(\boldsymbol{\Theta};\hat{\boldsymbol{\Theta}}\right)=\sum_{j=1}^{n}\hat{z}_j\left(\log w-\frac{\log 2\pi}{2}-\frac{\log\sigma_1^2}{2}-\frac{\left(y_j-\mathbf{x}_j'\boldsymbol{\beta}_1\right)^2}{2\sigma_1^2}\right)$$
$$+\left(1-\hat{z}_j\right)\Bigg(\log\left(1-w\right)-\frac{\log 2\pi}{2}-\frac{\log\sigma_2^2}{2}+\frac{\hat{u}_{2j}}{2}-\frac{\nu}{2}\hat{u}_{1j}$$
$$-\frac{\left(y_j-\mathbf{x}_j'\boldsymbol{\beta}_2\right)^2\hat{u}_{1j}}{2\sigma_2^2}-\log\Gamma\left(\frac{\nu}{2}\right)+\frac{\nu}{2}\log\left(\frac{\nu}{2}\right)+\left(\frac{\nu}{2}-1\right)\hat{u}_{2j}\Bigg). \tag{9}$$

Then, E- and M-steps of the EM algorithm will form the following iteratively reweighting algorithm.

Iteratively reweighting algorithm (EM algorithm)

1. Set initial parameter estimate $\boldsymbol{\Theta}^{(0)}$ and a stopping rule Δ.
2. Calculate the conditional expectations $\hat{z}_j^{(k)}$, $\hat{u}_{1j}^{(k)}$ and $\hat{u}_{2j}^{(k)}$ for the $(k+1)th$ for $k = 0, 1, 2, \ldots$ iteration using the Eqs. (36), (37) and (38) given in appendix.
3. Insert the current values $\hat{z}_j^{(k)}$, $\hat{u}_{1j}^{(k)}$, $\hat{u}_{2j}^{(k)}$ and $\hat{\boldsymbol{\Theta}}^{(k)}$ in $Q(\boldsymbol{\Theta}; \hat{\boldsymbol{\Theta}})$ to form $Q(\boldsymbol{\Theta}; \hat{\boldsymbol{\Theta}}^{(k)})$ and maximize $Q(\boldsymbol{\Theta}; \hat{\boldsymbol{\Theta}}^{(k)})$ with respect to the parameters $(w, \boldsymbol{\beta}_1, \sigma_1^2, \boldsymbol{\beta}_2, \sigma_2^2, \nu)$ to get new estimates for the parameters. This maximization will give the following updating equations:

$$\hat{w}^{(k+1)}=\frac{\sum_{j=1}^{n}\hat{z}_j^{(k)}}{n}, \tag{10}$$

$$\hat{\boldsymbol{\beta}}_1^{(k+1)}=\left(\sum_{j=1}^{n}\hat{z}_j^{(k)}\mathbf{x}_j\mathbf{x}_j'\right)^{-1}\left(\sum_{j=1}^{n}\hat{z}_j^{(k)}\mathbf{x}_jy_j\right), \tag{11}$$

$$\hat{\sigma}_1^{2(k+1)} = \frac{\sum_{j=1}^{n} \hat{z}_j^{(k)} \left(y_j - \mathbf{x}_j' \hat{\boldsymbol{\beta}}_1^{(k)}\right)^2}{\sum_{j=1}^{n} \hat{z}_j^{(k)}}, \tag{12}$$

$$\hat{\boldsymbol{\beta}}_2^{(k+1)} = \left(\sum_{j=1}^{n} \left(1 - \hat{z}_j^{(k)}\right) \hat{u}_{1j}^{(k)} \mathbf{x}_j \mathbf{x}_j'\right)^{-1} \left(\sum_{j=1}^{n} \left(1 - \hat{z}_j^{(k)}\right) \hat{u}_{1j}^{(k)} \mathbf{x}_j y_j\right), \tag{13}$$

$$\hat{\sigma}_2^{2(k+1)} = \frac{\sum_{j=1}^{n} \left(1 - \hat{z}_j^{(k)}\right) \hat{u}_{1j}^{(k)} \left(y_j - \mathbf{x}_j' \hat{\boldsymbol{\beta}}_2^{(k)}\right)^2}{\sum_{j=1}^{n} \left(1 - \hat{z}_j^{(k)}\right)}. \tag{14}$$

4. To obtain the $\hat{\nu}^{(k+1)}$ solve the following equation

$$\sum_{j=1}^{n} \left(1 - \hat{z}_j^{(k)}\right) \left(DG\left(\frac{\nu}{2}\right) - \log\left(\frac{\nu}{2}\right) - 1 - \hat{u}_{2j}^{(k)} + \hat{u}_{1j}^{(k)}\right) = 0, \tag{15}$$

   where $DG(\frac{\nu}{2}) = \frac{\Gamma'(\frac{\nu}{2})}{\Gamma(\frac{\nu}{2})}$ is the digamma function.

5. Repeat E and M steps until the convergence rule $\|\hat{\boldsymbol{\Theta}}^{(k+1)} - \hat{\boldsymbol{\Theta}}^{(k)}\| < \Delta$ is satisfied.

   Note that the Eq. (15) can be solved by using some numerical methods.

2.2 *Mixture Regression Estimation Based on the Mixture of Skew t (ST) and Skew Normal (SN) Distributions*

Next we will consider the parameter estimation for the mixture regression model assuming that the error terms have mixture of skew t and skew normal distributions. By taking this mixture of two different skew distributions we attempt to model skewness, as well as, the heavy-tailedness in the sub groups of the data.

For two-component mixture regression model given in (1), the conditional density of y given $\mathbf{x}$ is

$$f(y_j; \mathbf{x}_j, \boldsymbol{\Theta}) = w f_{ST}\left(y_j; \mathbf{x}_j' \boldsymbol{\beta}_1, \sigma_1^2, \lambda_1, \nu\right) + (1 - w) f_{SN}\left(y_j; \mathbf{x}_j' \boldsymbol{\beta}_2, \sigma_2^2, \lambda_2\right), \tag{16}$$

where $f_{ST}(\cdot)$ is the density function of the skew t distribution proposed by Azzalini and Capitaino (2003) with the parameters $(\sigma_1^2, \lambda_1, \nu)$ and $f_{SN}(\cdot)$ is the density function of the skew normal distribution proposed by Azzalini (1985, 1986) with the parameters (σ_2^2, λ_2). Note that the skew t is the distribution of ϵ_1 and the skew normal is the distribution of ϵ_2. Let $\boldsymbol{\Theta} = (w, \boldsymbol{\beta}_1, \sigma_1^2, \lambda_1, \nu, \boldsymbol{\beta}_2, \sigma_2^2, \lambda_2)'$ be the unknown parameter vector for this model. Notice that we have extra two skewness parameters to be estimated compare to the model given in Sect. 2.1. In this mixture regression model we note that different from the symmetric case $E(\epsilon) \neq 0$. Therefore, when we estimate the intercept we take into consideration $\widehat{E(\epsilon)}$.

To find the ML estimator of the unknown parameter $\boldsymbol{\Theta}$ we should maximize the following log-likelihood function

$$\ell(\boldsymbol{\Theta}) = \sum_{j=1}^{n} \log \left(w f_{ST}(y_j; \mathbf{x}_j' \boldsymbol{\beta}_1, \sigma_1^2, \lambda_1, \nu) + (1-w) f_{SN}(y_j; \mathbf{x}_j' \boldsymbol{\beta}_2, \sigma_2^2, \lambda_2) \right). \tag{17}$$

Since the log-likelihood function does not have an explicit maximizer, the estimates for the unknown parameter vector $\boldsymbol{\Theta}$ can be again obtained by using the EM algorithm.

Let z_j define as in Eq. (4), for $j = 1, \ldots, n$. The joint density function of two-component mixture regression model is

$$f(y_j, z_j) = \left(w f_{ST}(y_j; \mathbf{x}_j' \boldsymbol{\beta}_1, \sigma_1^2, \lambda_1, \nu) \right)^{z_j} \left((1-w) f_{SN}(y_j; \mathbf{x}_j' \boldsymbol{\beta}_2, \sigma_2^2, \lambda_2) \right)^{1-z_j}. \tag{18}$$

To represent this joint density function in terms of the normal distribution, we will use the stochastic representation of the skew t and the skew normal distributions. By doing this we will simplify the steps of the EM algorithm. One can see the papers proposed by Azzalini and Capitaino (2003), Azzalini (1986, p. 201), Henze (1986, Theorem 1) to get more details for the stochastic representation of the skew t and the skew normal distributions. Using the scale mixture representation of the skew t distribution and the stochastic representation of the skew t and the skew normal distribution following conditional distributions can be given (Lin et al. 2007; Liu and Lin 2014). Let γ and τ be the latent variables. Then, we have

$$\begin{aligned}
y_j | \gamma_j, \tau_j &\sim N\left(\mathbf{x}_j' \boldsymbol{\beta}_1 + \alpha_1 \gamma_j, \frac{\kappa_1^2}{\tau_j} \right), \\
y_j | \tau_j &\sim TN\left(0, \frac{1}{\tau_j}; (0, \infty) \right), \ \tau_j \sim Gamma\left(\frac{\nu}{2}, \frac{\nu}{2} \right), \\
y_j | \gamma_j &\sim N\left(\mathbf{x}_j' \boldsymbol{\beta}_2 + \alpha_2 \gamma_j, \kappa_2^2 \right), \ \gamma_j \sim TN\left(0, 1; (0, \infty)\right),
\end{aligned}$$

where $\delta_{\lambda_1} = \lambda_1 / \sqrt{1+\lambda_1^2}$, $\delta_{\lambda_2} = \lambda_2 / \sqrt{1+\lambda_2^2}$, $\alpha_1 = \sigma_1 \delta_{\lambda_1}$, $\alpha_2 = \sigma_2 \delta_{\lambda_2}$, $\kappa_1^2 = \sigma_1^2(1 - \delta_{\lambda_1}^2)$, $\kappa_2^2 = \sigma_2^2(1 - \delta_{\lambda_2}^2)$ and $TN(\cdot)$ shows the truncated normal distribution.

Using the conditional distributions given above the joint density function given in (18) can be rewritten as

$$f(y_j,\gamma_j,\tau_j,z_j)=\left(w\frac{(\nu/2)^{\nu/2}\tau_j^{\nu/2}}{\pi\Gamma\left(\frac{\nu}{2}\right)\sqrt{\kappa_1^2}}e^{-\frac{\nu\tau_j}{2}-\frac{\tau_j(y_j-\mathbf{x}_j^{'}\boldsymbol{\beta}_1-\alpha_1\gamma_j)^2}{2\kappa_1^2}-\frac{\tau_j\gamma_j^2}{2}}\right)^{z_j}$$
$$\left(\frac{(1-w)}{\pi\sqrt{\kappa_2^2}}e^{-\frac{\left(y_j-\mathbf{x}_j^{'}\boldsymbol{\beta}_2-\alpha\gamma_j\right)^2}{2\kappa_2^2}-\frac{\gamma_j^2}{2}}\right)^{1-z_j}. \quad (19)$$

Note that in this model $(\boldsymbol{\gamma},\boldsymbol{\tau},\mathbf{z})$ will be regarded as the missing and $\mathbf{y}$ will be the observed data, where $\mathbf{y}=(y_1,\ldots,y_n)$, $\boldsymbol{\gamma}=(\gamma_1,\ldots,\gamma_n)$, $\boldsymbol{\tau}=(\tau_1,\ldots,\tau_n)$ and $\mathbf{z}=(z_1,\ldots,z_n)$. Let $(\mathbf{y},\boldsymbol{\gamma},\boldsymbol{\tau},\mathbf{z})$ be the complete data. Then, using the complete data joint density function given in (19), the complete data log-likelihood function can be obtained as follows

$$\ell_c(\boldsymbol{\Theta};\mathbf{y},\boldsymbol{\gamma},\boldsymbol{\tau},\mathbf{z})=\sum_{j=1}^{n}z_j\left(\log w-\log\pi-\frac{\log\kappa_1^2}{2}+\frac{\nu}{2}\log\left(\frac{\nu}{2}\right)-\log\left(\Gamma\left(\frac{\nu}{2}\right)\right)\right.$$
$$\left.+\frac{\nu}{2}\log\tau_j-\frac{\nu\tau_j}{2}-\frac{(y_j-\mathbf{x}_j^{'}\boldsymbol{\beta}_1-\alpha_1\gamma_j)^2}{2\kappa_1^2/\tau_j}-\frac{\tau_j\gamma_j^2}{2}\right)+(1-z_j)$$
$$\left(\log(1-w)-\log\pi-\frac{\log\kappa_2^2}{2}-\frac{(y_j-\mathbf{x}_j^{'}\boldsymbol{\beta}_2-\alpha_2\gamma_j)^2}{2\kappa_2^2}-\frac{\gamma_j^2}{2}\right). \quad (20)$$

Since we cannot be able to observe the missing data $(\boldsymbol{\gamma},\boldsymbol{\tau},\mathbf{z})$ this complete data log-likelihood function cannot be used to obtain the estimator for $\boldsymbol{\Theta}$. To overcome this problem we have to take the conditional expectation of the complete data log-likelihood function given the observed data $\mathbf{y}$. This will be the E-step of the EM algorithm

E-step

$$E\left(\ell_c(\boldsymbol{\Theta};\mathbf{y},\boldsymbol{\gamma},\boldsymbol{\tau},\mathbf{z})|y_j\right)=\sum_{j=1}^{n}E(z_j|y_j)\left(\log w-\frac{\log\kappa_1^2}{2}+\frac{\nu}{2}\log\left(\frac{\nu}{2}\right)-\log\Gamma\left(\frac{\nu}{2}\right)\right)$$
$$+\frac{\nu E(z_j\log\tau_j|y_j)}{2}-\frac{\nu E(Z_j\tau_j|y_j)}{2}-\frac{E(Z_j\tau_j|y_j)\left(y_j-\mathbf{x}_j^{'}\boldsymbol{\beta}_1\right)^2}{2\kappa_1^2}$$
$$-\frac{\alpha_1^2E(z_j\tau_j\gamma_j^2|y_j)}{2\kappa_1^2}+\frac{\alpha_1E(Z_j\tau_j\gamma_j|y_j)(y_j-\mathbf{x}_j^{'}\boldsymbol{\beta}_1)}{\kappa_1^2}$$
$$+(1-E(z_j|y_j))\left(\log(1-w)-\frac{\log\kappa_2^2}{2}-\frac{\left(y_j-\mathbf{x}_j^{'}\boldsymbol{\beta}_2\right)^2}{2\kappa_2^2}\right.$$
$$\left.+\frac{\alpha_2E(\gamma_j|y_j)(y_j-\mathbf{x}_j^{'}\boldsymbol{\beta}_2)}{\kappa_2^2}-\frac{\alpha_2^2E(\gamma_j^2|y_j)}{2\kappa_2^2}\right). \quad (21)$$

To obtain the conditional expectation of the complete data log-likelihood function we have to find $\hat{z}_j=E(z_j|y_j,\hat{\boldsymbol{\Theta}})$, $\hat{s}_{1j}=E(z_j\tau_j|y_j,\hat{\boldsymbol{\Theta}})$, $\hat{s}_{2j}=E(z_j\gamma_j\tau_j|y_j,\hat{\boldsymbol{\Theta}})$,

$\hat{s}_{3j} = E(z_j\gamma_j^2\tau_j|y_j,\hat{\boldsymbol{\Theta}})$, $\hat{s}_{4j} = E(z_j\log(\tau_j)|y_j,\hat{\boldsymbol{\Theta}})$, $\hat{t}_{1j} = E(\gamma_j|y_j,\hat{\boldsymbol{\Theta}})$ and $\hat{t}_{2j} = E(\gamma_j^2|y_j,\hat{\boldsymbol{\Theta}})$ given in (39)–(45).

M-step: For the M step of the EM algorithm, the expected complete data log-likelihood function will be maximized with respect to the parameter $\boldsymbol{\Theta}$

$$Q\left(\boldsymbol{\Theta};\hat{\boldsymbol{\Theta}}\right) = \sum_{j=1}^{n}\hat{z}_j\left(\log w - \frac{1}{2}\log(\kappa_1^2) + \frac{\nu}{2}\log\left(\frac{\nu}{2}\right) - \log\left(\Gamma\left(\frac{\nu}{2}\right)\right)\right) - \frac{\nu\hat{s}_{1j}}{2}$$
$$+\frac{\nu\hat{s}_{4j}}{2} - \frac{\hat{s}_{1j}(y_j-\mathbf{x}_j'\boldsymbol{\beta}_1)^2}{2\kappa_1^2} + \frac{\alpha_1\hat{s}_{2j}(y_j-\mathbf{x}_j'\boldsymbol{\beta}_1)}{\kappa_1^2} - \frac{\alpha_1^2\hat{s}_{3j}}{2\kappa_1^2} + (1-\hat{z}_j)$$
$$\left(\log(1-w) - \frac{\log\kappa_2^2}{2} - \frac{(y_j-\mathbf{x}_j'\boldsymbol{\beta}_2)^2 - 2\alpha_2\hat{t}_{1j}(y_j-\mathbf{x}_j'\boldsymbol{\beta}_2) + \alpha_2^2\hat{t}_{2j}}{2\kappa_2^2}\right). \tag{22}$$

Similar to the iteratively reweighting algorithm given in Sect. 2.1, we can give the following algorithm based on the steps of the EM algorithm for the two-component mixture regression model obtained from the skew t and skew normal distributions.

Iteratively reweighting algorithm (EM algorithm)

1. Set an initial parameter estimates $\boldsymbol{\Theta}^{(0)}$ and stopping rule Δ.
2. Use $\hat{\boldsymbol{\Theta}}^{(k)}$ to compute the conditional expectations $\hat{z}_j^{(k)}, \hat{s}_{1j}^{(k)}, \hat{s}_{2j}^{(k)}, \hat{s}_{3j}^{(k)}, \hat{s}_{4j}^{(k)}, \hat{t}_{1j}^{(k)}, \hat{t}_{2j}^{(k)}$ for $k = 0, 1, 2, \ldots$ from the Eqs. (39)–(45) given in appendix.
3. Insert $\hat{z}_j^{(k)}, \hat{s}_{1j}^{(k)}, \hat{s}_{2j}^{(k)}, \hat{s}_{3j}^{(k)}, \hat{s}_{4j}^{(k)}, \hat{t}_{1j}^{(k)}, \hat{t}_{2j}^{(k)}$ and $\hat{\boldsymbol{\Theta}}^{(k)}$ in $Q(\boldsymbol{\Theta};\hat{\boldsymbol{\Theta}})$ to form $Q(\boldsymbol{\Theta};\hat{\boldsymbol{\Theta}}^{(k)})$.

   Maximize the function $Q(\boldsymbol{\Theta};\hat{\boldsymbol{\Theta}}^{(k)})$ given in (22) with respect to the parameters $(w, \boldsymbol{\beta}_1, \sigma_1^2, \lambda_1, \boldsymbol{\beta}_2, \sigma_2^2, \lambda_2)$ to get $(k+1)$ iterated values

$$\hat{w}^{(k+1)} = \frac{\sum_{j=1}^{n}\hat{z}_j^{(k)}}{n}, \tag{23}$$

$$\hat{\boldsymbol{\beta}}_1^{(k+1)} = \left(\sum_{j=1}^{n}\hat{s}_{1j}^{(k)}\mathbf{x}_j\mathbf{x}_j'\right)^{-1}\left(\sum_{j=1}^{n}\left(y_j\hat{s}_{1j}^{(k)} - \hat{\delta}_{\lambda_1}^{(k)}\hat{s}_{2j}^{(k)}\right)\mathbf{x}_j\right), \tag{24}$$

$$\hat{\alpha}_1^{(k+1)} = \frac{\sum_{j=1}^{n}\hat{s}_{2j}^{(k)}(y_j-\mathbf{x}_j'\hat{\boldsymbol{\beta}}_1^{(k)})}{\sum_{j=1}^{n}\hat{s}_{3j}^{(k)}}, \tag{25}$$

$$\hat{\kappa}_1^{2(k+1)} = \frac{\sum_{j=1}^{n}\hat{s}_{1j}^{(k)}(y_j-\mathbf{x}_j'\hat{\boldsymbol{\beta}}_1^{(k)})^2 - 2\hat{\alpha}_1^{(k)}\hat{s}_{2j}^{(k)}(y_j-\mathbf{x}_j'\hat{\boldsymbol{\beta}}_1^{(k)}) + \hat{\alpha}_1^{2(k)}\hat{s}_{2j}^{(k)}}{\sum_{j=1}^{n}\hat{z}_j^{(k)}}, \tag{26}$$

$$\hat{\boldsymbol{\beta}}_2^{(k+1)} = \left(\sum_{j=1}^{n} \left(1 - \hat{z}_j^{(k)}\right) \mathbf{x}_j \mathbf{x}_j^{'} \right)^{-1} \left(\sum_{j=1}^{n} \left(1 - \hat{z}_j^{(k)}\right) \left(y_j - \hat{\alpha}_2^{(k)} \hat{t}_{1j}^{(k)}\right) \mathbf{x}_j \right), \quad (27)$$

$$\hat{\alpha}_2^{(k+1)} = \frac{\sum_{j=1}^{n} \left(1 - \hat{z}_j^{(k)}\right) \hat{t}_{1j}^{(k)} (y_j - \mathbf{x}_j^{'} \hat{\boldsymbol{\beta}}_2^{(k)})}{\sum_{j=1}^{n} \left(1 - \hat{z}_j^{(k)}\right) \hat{t}_{2j}^{(k)}}, \quad (28)$$

$$\hat{\kappa}_2^{2(k+1)} = \frac{1}{\sum_{j=1}^{n} \left(1 - \hat{z}_j^{(k)}\right)} \sum_{j=1}^{n} \left(1 - \hat{z}_j^{(k)}\right) \Big((y_j - \mathbf{x}_j^{'} \hat{\boldsymbol{\beta}}_2^{(k)})^2 - 2\hat{\alpha}_2^{(k)} \hat{t}_{1j}^{(k)} (y_j - \mathbf{x}_j^{'} \hat{\boldsymbol{\beta}}_2^{(k)}) + \hat{\alpha}_2^{2(k)} \hat{t}_{2j}^{(k)} \Big). \quad (29)$$

Then, we obtain the $\hat{\sigma}_1^{2(k+1)}$, $\hat{\lambda}_1^{(k+1)}$, $\hat{\sigma}_2^{2(k+1)}$ and $\hat{\lambda}_2^{(k+1)}$ parameter estimates

$$\hat{\sigma}_1^{2(k+1)} = \hat{\kappa}_1^{2(k+1)} + \hat{\alpha}_1^{2(k+1)}, \quad (30)$$

$$\hat{\lambda}_1^{(k+1)} = \hat{\delta}_{\lambda_1}^{(k+1)} \left(1 - \hat{\delta}_{\lambda_1}^{2(k+1)}\right)^{-1/2}, \quad (31)$$

$$\hat{\sigma}_2^{2(k+1)} = \hat{\kappa_2}^{2(k+1)} + \hat{\alpha_2}^{2(k+1)}, \quad (32)$$

$$\hat{\lambda}_2^{(k+1)} = \hat{\delta}_{\lambda_2}^{(k+1)} \left(1 - \hat{\delta}_{\lambda_2}^{2(k+1)}\right)^{-1/2}, \quad (33)$$

where $\hat{\delta}_{\lambda_1}^{(k+1)} = \hat{\alpha_1}^{(k+1)} / \hat{\sigma}_1^{(k+1)}$ and $\hat{\delta}_{\lambda_2}^{(k+1)} = \hat{\alpha_2}^{(k+1)} / \hat{\sigma}_2^{(k+1)}$.

4. Also $(k+1)$th value of λ_1 can be found by solving following equation

$$\delta_{\lambda_1}(1 - \delta_{\lambda_1}^2) \sum_{j=1}^{n} \hat{z}_j^{(k)} - \delta_{\lambda_1} \left(\sum_{j=1}^{n} \hat{s}_{1j}^{(k)} \frac{(y_j - \mathbf{x}_j' \hat{\boldsymbol{\beta}}_1^{(k)})^2}{\hat{\sigma}_1^{2(k)}} + \sum_{j=1}^{n} \hat{s}_{3j}^{(k)} \right) + (1 + \delta_{\lambda_1}^2) \sum_{j=1}^{n} \hat{s}_{2j}^{(k)} \frac{(y_j - \mathbf{x}_j' \hat{\boldsymbol{\beta}}_1^{(k)})}{\hat{\sigma}_1^{(k)}} = 0. \quad (34)$$

The $(k+1)$th values of ν can be calculated solving the following equation

$$\log\left(\frac{\nu}{2}\right) + 1 - DG\left(\frac{\nu}{2}\right) + \frac{\sum_{j=1}^{n} \left(\hat{s}_{4j}^{(k)} - \hat{s}_{1j}^{(k)}\right)}{\sum_{j=1}^{n} \hat{z}_j^{(k)}} = 0. \quad (35)$$

5. Repeat E and M steps until the convergence rule $\|\hat{\boldsymbol{\Theta}}^{(k+1)} - \hat{\boldsymbol{\Theta}}^{(k)}\| < \Delta$ is satisfied.

Note that we can solve the Eqs. (34) and (35) using some numerical algorithms.

3 Simulation Study

In this section we will give a simulation study to assess and compare the performances of the mixture regression estimators proposed in this paper with the existing mixture regression estimators in the literature. We specifically compare the mixture regression estimators obtained from normal and t (MixregNt) distributions with the estimators obtained from normal (MixregN) and t (Mixregt) distributions for the two-component mixture regression models for the symmetric case. For the skew case, we compare the mixture regression estimators obtained from the skew t and the skew normal (MixregSTSN) distributions with the estimators obtained from skew normal (MixregSN) and skew t (MixregST) distributions for the two-component mixture regression models. The compression will be done in terms of bias and mean square error (MSE) which are given the following formulas

$$\widehat{bias}(\hat{\theta}) = \bar{\theta} - \theta,\ \widehat{MSE}(\hat{\theta}) = \frac{1}{N}\sum_{i=1}^{N}(\hat{\theta}_i - \theta)^2,$$

where θ is the true parameter value, $\hat{\theta}_i$ is the ith simulated parameter estimate, $\bar{\theta} = \frac{1}{N}\sum_{i=1}^{N}\hat{\theta}_i$ and $N = 500$ is the replication number. For the sample sizes, we take $n = 200$ and $n = 400$. The simulation is conducted using $MATLAB\ R2013a$. The $MATLAB$ codes can be obtained upon request.

Alternatively, the MSE for the $\hat{\boldsymbol{\Theta}}$ which can be defined as $\|\hat{\boldsymbol{\Theta}} - \boldsymbol{\Theta}_0\|^2$, where $\boldsymbol{\Theta}_0$ is the true parameter, can be also used to illustrate the performance of the parameter vector as is suggested by one of the referee. However, to see the performance of each parameter we prefer computing the MSE for each parameter separately. We compare the both the MSE values and observe the similar behavior.

The data $\{(x_{1j}, x_{2j}, y_j), j = 1, \ldots, n\}$ are generated from the following two-component mixture regression model (Bai et al. 2012)

$$Y = \begin{cases} 0 + X_1 + X_2 + \epsilon_1, Z = 1, \\ 0 - X_1 - X_2 + \epsilon_2, Z = 2, \end{cases}$$

where $P(Z = 1) = 0.25 = w_1$, $X_1 \sim N(0, 1)$ and $X_2 \sim N(0, 1)$. The values of the regression coefficients are $\boldsymbol{\beta}_1 = (\beta_{10}, \beta_{11}, \beta_{12})' = (0, 1, 1)'$ and $\boldsymbol{\beta}_2 = (\beta_{20}, \beta_{21}, \beta_{22})' = (0, -1, -1)'$, respectively.

We consider the following error distributions for the symmetric (i) and skew (ii) cases.

(i) Case I: $\epsilon_1, \epsilon_2 \sim N(0, 1)$, the standard normal distribution.
Case II: $\epsilon_1, \epsilon_2 \sim t_3(0, 1)$, the t distribution with 3 degrees of freedom.
Case III: $\epsilon_1 \sim N(0, 1)$ and $\epsilon_2 \sim t_3(0, 1)$.
Case IV: $\epsilon_1, \epsilon_2 \sim N(0, 1)$ and we also added %5 outliers at $X_1 = 20$, $X_2 = 20$ and $Y = 100$.
(ii) Case I: $\epsilon_1, \epsilon_2 \sim SN(0, 1, 0.5)$, the skew normal distribution.
Case II: $\epsilon_1, \epsilon_2 \sim ST(0, 1, 0.5, 3)$, the skew t distribution with 3 degrees of freedom.
Case III: $\epsilon_1 \sim ST(0, 1, 0.5, 3)$ and $\epsilon_2 \sim SN(0, 1, 0.5)$.
Case IV: $\epsilon_1, \epsilon_2 \sim N(0, 1)$ and we also added %5 outliers at $X_1 = 20$, $X_2 = 20$ and $Y = 100$.

The simulation results are summarized in Tables 1, 2, 3 and 4. Tables 1 and 2 show the simulation results for the estimators based on MixregNt with the error distributions given in case (i). For the Case I the best result is obtained from the estimators based on MixregN. For this case, the estimators based on Mixregt and the estimators based on MixregNt have similar behavior. For the error distribution given in Case II the best behavior is obtained, as expected, from Mixregt. In this case, the estimators based on MixregN are drastically affected. The proposed estimators (MixregNt) again have similar behavior with the estimators obtained from Mixregt which shows that it tolerates the heavy-tailedness. The estimators obtained from MixregNt perform the best for the error distribution given in Case III. In this case the estimator obtained from MixregN again has the worst performance. On the other hand, the performance of the estimators based on Mixregt is comparable with the estimators based on MixregNt. Finally, for the outlier case (Case IV) the behavior of the estimators based on MixregN and Mixregt is very similar. In both cases the worst performance is obtained for small groups. That is, they fail to find the regression line for the smaller group. In contrast, the estimators based on MixregNt can be able to accommodate the regression lines for both groups. This can be seen from the smaller bias and the MSE values. In summary, for all the cases considered in this part of the simulation the behavior of the proposed estimators is comparable with the counterparts.

In Tables 3 and 4 we summarize the simulation results obtained from the skew distributions with the error distributions given in case (ii). From this table we can observe that when the error distribution is the mixture of skew normal distribution, the estimators obtained from MixregSN behave better than the other cases. The same behavior can be noticed for the skew t distribution as well. When the error distribution is the mixture of the skew t and the skew normal the estimators obtained from MixregSTSN outperform the counterparts in terms of the MSE values. In this case, the estimators based on MixregSN have the worst performance. When we add

Table 1 MSE (bias) values of estimates for $n = 200$

	MixregN	Mixregt	MixregNt
Case I: $\epsilon_1, \epsilon_2 \sim N(0, 1)$			
β_{10}:0	0.0468 (−0.0039)	0.0547 (−0.0038)	0.0468 (−0.0036)
β_{20}:0	0.0099 (0.0028)	0.0112 (0.0056)	0.0111 (0.0048)
β_{11}:1	0.0457 (−0.0273)	0.0483 (−0.0364)	0.0529 (−0.0573)
β_{21}:−1	0.0147 (0.0055)	0.0101 (0.0090)	0.0096 (−0.0025)
β_{12}:1	0.0414 (−0.0015)	0.0463 (−0.0046)	0.0545 (−0.0353)
β_{22}:−1	0.0216 (0.0026)	0.0108 (0.0013)	0.0109 (−0.0090)
w:0.25	0.0029 (0.0071)	0.0022 (0.0048)	0.0036 (0.0335)
Case II: $\epsilon_1, \epsilon_2 \sim t_3(0, 1)$			
β_{10}:0	12.7323 (0.2016)	0.0846 (0.0245)	0.1378 (0.0265)
β_{20}:0	1.9712 (−0.0228)	0.0146 (−0.0048)	0.0140 (−0.0047)
β_{11}:1	10.6545 (0.3274)	0.1037 (−0.0479)	0.1417 (0.0332)
β_{21}:−1	1.5718 (−0.0867)	0.0128 (−0.0110)	0.0136 (0.0148)
β_{12}:1	8.2033 (0.0267)	0.0806 (0.0024)	0.1288 (0.0403)
β_{22}:−1	2.8037 (0.2205)	0.0143 (−0.0149)	0.0156 (0.0126)
w:0.25	0.0250 (−0.0347)	0.0030 (0.0062)	0.0045 (−0.0374)
Case III: $\epsilon_1 \sim N(0, 1)$ and $\epsilon_2 \sim t_3(0, 1)$			
β_{10}:0	6.5822 (0.0716)	0.0608 (−0.0015)	0.0564 (0.0039)
β_{20}:0	0.3372 (−0.0249)	0.0144 (0.0036)	0.0139 (0.0030)
β_{11}:1	5.0332 (0.0836)	0.0494 (−0.0387)	0.0459 (−0.0170)
β_{21}:−1	0.2761 (0.0073)	0.0138 (−0.0185)	0.0133 (−0.0105)
β_{12}:1	5.1167 (0.1571)	0.0673 (−0.0496)	0.0647 (−0.0287)
β_{22}:−1	0.5948 (0.0062)	0.0136 (−0.0213)	0.0134 (−0.0145)
w:0.25	0.0120 (0.002)	0.0039 (0.0352)	0.0029 (0.0174)
Case IV: $\epsilon_1, \epsilon_2 \sim N(0, 1)$ (% 5 outliers)			
β_{10}:0	2.3909 (0.1047)	1.4186 (0.0618)	0.0506 (0.0003)
β_{20}:0	0.0149 (0.0029)	0.0110 (0.0022)	0.0173 (0.0036)
β_{11}:1	3.1304 (1.4785)	2.7396 (1.4819)	0.1498 (−0.0521)
β_{21}:−1	0.0799 (0.2483)	0.0241 (0.1065)	0.2815 (0.1407)
β_{12}:1	3.2398 (1.5125)	2.8212 (1.5088)	0.1734 (−0.0633)
β_{22}:−1	0.0834 (0.2492)	0.0246 (0.1078)	0.1636 (0.1285)
w:0.25	0.0095 (−0.0943)	0.0062 (−0.0753)	0.0081 (−0.0208)

Note Value in parentheses indicates the bias

the leverage point (the error distribution given in Case IV) the behavior of all the estimators are similarly worse. However, the estimators obtained from MixregST and MixregSTSN give comparable results which have smaller bias and MSE than MixregSN.

Table 2 MSE (bias) values of estimates for $n = 400$

	MixregN	Mixregt	MixregNt
Case I: $\epsilon_1, \epsilon_2 \sim N(0, 1)$			
β_{10}:0	0.0202 (−0.0021)	0.0260 (−0.0004)	0.0217 (−0.0003)
β_{20}:0	0.0041 (0.0018)	0.0052 (0.0036)	0.0050 (0.0031)
β_{11}:1	0.0160 (0.0033)	0.0206 (−0.0062)	0.0188 (−0.0199)
β_{21}:−1	0.0045 (−0.0035)	0.0053 (0.0018)	0.0051 (−0.0069)
β_{12}:1	0.0177 (−0.0019)	0.0254 (−0.0099)	0.0210 (−0.0213)
β_{22}:−1	0.0038 (−0.0038)	0.0049 (0.0035)	0.0046 (−0.0055)
w:0.25	0.0010 (0.0037)	0.0011 (0.0007)	0.0018 (0.0241)
Case II: $\epsilon_1, \epsilon_2 \sim t_3(0, 1)$			
β_{10}:0	13.4210 (−0.1318)	0.0376 (−0.0066)	0.0473 (−0.0154)
β_{20}:0	1.4586 (0.0306)	0.0068 (−0.0021)	0.0066 (−0.0026)
β_{11}:1	9.2787 (0.5967)	0.0335 (0.0019)	0.0449 (0.0632)
β_{21}:−1	1.8295 (0.0194)	0.0063 (0.0015)	0.0079 (0.0314)
β_{12}:1	11.8714 (0.4395)	0.0388 (−0.0008)	0.0533 (0.0722)
β_{22}:−1	0.7543 (0.01106)	0.0064 (0.0024)	0.0082 (0.0308)
w:0.25	0.0171 (−0.0596)	0.0014 (0.0040)	0.0037 (−0.0454)
Case III: $\epsilon_1 \sim N(0, 1)$ and $\epsilon_2 \sim t_3(0, 1)$			
β_{10}:0	7.7436 (0.0036)	0.0247 (0.0066)	0.0240 (0.0060)
β_{20}:0	0.3984 (0.0218)	0.0070 (0.0050)	0.0068 (0.0050)
β_{11}:1	5.8227 (0.1401)	0.0251 (−0.0372)	0.0206 (−0.0124)
β_{21}:−1	0.4005 (−0.0086)	0.0062 (−0.0126)	0.0060 (−0.0039)
β_{12}:1	6.6747 (0.2501)	0.0244 (−0.0365)	0.0213 (−0.0125)
β_{22}:−1	0.3341 (−0.0023)	0.0064 (−0.0143)	0.0063 (−0.0060)
w:0.25	0.0090 (−0.0070)	0.0021 (0.0289)	0.0015 (0.0077)
Case IV: $\epsilon_1, \epsilon_2 \sim N(0, 1)$ (% 10 outliers)			
β_{10}:0	1.4233 (0.1705)	1.0180 (0.2007)	0.0236 (0.0089)
β_{20}:0	0.0079 (0.0102)	0.0048 (0.0045)	0.0045 (0.0042)
β_{11}:1	2.9197 (1.5268)	2.5645 (1.4732)	0.0280 (−0.0003)
β_{21}:−1	0.0774 (0.2583)	0.0162 (0.1023)	0.0450 (0.0842)
β_{12}:1	2.7131 (1.4602)	2.6908 (1.5157)	0.0277 (0.0093)
β_{22}:−1	0.0759 (0.2536)	0.0150 (0.0961)	0.0264 (0.0733)
w:0.25	0.0099 (−0.0981)	0.0069 (−0.0817)	0.0036 (−0.0412)

Note Value in parentheses indicates the bias

Note that from the computational point of view, computing the estimators based on MixregSTSN is less intensive than the estimators obtained from MixregST. Therefore, even they show similar behavior MixregSTSN should be preferred.

Table 3 MSE (bias) values of estimates for $n = 200$

	MixregSN	MixregST	MixregSTSN
Case I: $\epsilon_1, \epsilon_2 \sim SN(0, 1, 0.5)$			
β_{10}:0	0.0396 (0.0145)	0.3796 (0.5558)	0.2090 (0.3916)
β_{20}:0	0.0083 (0.0124)	0.0388 (0.1602)	0.0095 (0.0228)
β_{11}:1	0.0322 (−0.0016)	0.0481 (−0.0041)	0.0445 (−0.0325)
β_{21}:−1	0.0080 (−0.0025)	0.0107 (0.0166)	0.0083 (−0.0082)
β_{12}:1	0.0366 (−0.0051)	0.0509 (−0.0150)	0.0491 (−0.0357)
β_{22}:−1	0.0080 (0.0019)	0.0104 (0.0212)	0.0083 (−0.0039)
w:0.25	0.0021 (0.0044)	0.0029 (−0.0087)	0.0028 (0.0162)
Case II: $\epsilon_1, \epsilon_2 \sim ST(0, 1, 0.5, 3)$			
β_{10}:0	5.5600 (0.3072)	0.8540 (0.7143)	0.4782 (0.2833)
β_{20}:0	1.7060 (−0.0335)	0.0239 (0.0805)	0.0421 (−0.0715)
β_{11}:1	5.1535 (0.2173)	0.1600 (0.0246)	0.2281 (−0.1303)
β_{21}:−1	0.9447 (0.0240)	0.0154 (0.0272)	0.0198 (−0.0147)
β_{12}:1	2.9528 (0.0176)	0.1445 (0.0178)	0.2302 (−0.1455)
β_{22}:−1	3.6893 (−0.0872)	0.0159 (0.0190)	0.0206 (−0.0084)
w:0.25	0.0175 (−0.0268)	0.0041 (−0.0128)	0.0084 (0.0444)
Case III $\epsilon_1 \sim ST(0, 1, 0.5, 3)$ and $\epsilon_2 \sim SN(0, 1, 0.5)$			
β_{10}:0	2.7217 (0.2228)	0.6258 (0.6722)	0.2932 (0.4207)
β_{20}:0	0.0555 (−0.0775)	0.0250 (0.1057)	0.0102 (−0.0123)
β_{11}:1	2.3168 (0.2017)	0.1033 (0.0327)	0.0921 (−0.0165)
β_{21}:−1	0.1975 (0.0372)	0.0108 (0.0305)	0.0088 (0.0059)
β_{12}:1	2.2086 (0.0959)	0.1158 (0.0228)	0.0958 (−0.0324)
β_{22}:−1	0.0244 (0.0598)	0.0111 (0.0314)	0.0090 (0.0079)
w:0.25	0.0079 (−0.0450)	0.0049 (−0.0369)	0.0031 (−0.0026)
Case IV: $\epsilon_1, \epsilon_2 \sim N(0, 1)$ (% 5 outliers)			
β_{10}:0	2.8415 (−0.5539)	5.7397 (2.1180)	1.7247 (0.5967)
β_{20}:0	0.2470 (−0.4804)	0.0437 (−0.1774)	0.1687 (−0.3970)
β_{11}:1	3.4886 (1.5398)	2.9568 (1.5342)	3.0075 (1.5224)
β_{21}:−1	0.0703 (0.2303)	0.0263 (0.1172)	0.0310 (0.1347)
β_{12}:1	3.2631 (1.4610)	2.4560 (1.3592)	2.7780 (1.4447)
β_{22}:−1	0.0784 (0.2423)	0.0278 (0.1211)	0.0347 (0.1435)
w:0.25	0.0093 (−0.0928)	0.0113 (−0.1031)	0.0049 (−0.0632)

Note Value in parentheses indicates the bias

Table 4 MSE (bias) values of estimates for $n = 400$

	MixregSN	MixregST	MixregSTSN
Case I: $\epsilon_1, \epsilon_2 \sim SN(0, 1, 0.5)$			
β_{10}:0	0.0172 (0.0055)	0.3194 (0.5421)	0.1741 (0.3920)
β_{20}:0	0.0035 (0.0040)	0.0258 (0.1422)	0.0038 (0.0147)
β_{11}:1	0.0145 (−0.0061)	0.0229 (−0.0045)	0.0197 (−0.0334)
β_{21}:−1	0.0035 (−0.0066)	0.0045 (0.0165)	0.0039 (−0.0122)
β_{12}:1	0.0145 (0.0012)	0.0220 (−0.0019)	0.0181 (−0.0235)
β_{22}:−1	0.0035 (−0.0004)	0.0048 (0.0194)	0.0036 (−0.0065)
w:0.25	0.0010 (0.0010)	0.0015 (−0.0147)	0.0013 (0.0126)
Case II: $\epsilon_1, \epsilon_2 \sim ST(0, 1, 0.5, 3)$			
β_{10}:0	9.1023 (0.5460)	0.6374 (0.6651)	0.2101 (0.3311)
β_{20}:0	0.7382 (−0.1361)	0.0105 (0.0402)	0.0141 (−0.0717)
β_{11}:1	7.2165 (0.2491)	0.0540 (0.0211)	0.0783 (−0.0925)
β_{21}:−1	0.7558 (0.0446)	0.0078 (0.0307)	0.0091 (−0.0188)
β_{12}:1	7.3711 (0.2215)	0.1714 (0.0671)	0.0785 (−0.0720)
β_{22}:−1	1.9910 (−0.0097)	0.0078 (0.0331)	0.0098 (−0.0157)
w:0.25	0.0144 (−0.0534)	0.0025 (−0.0217)	0.0044 (0.0441)
Case III: $\epsilon_1 \sim ST(0, 1, 0.5, 3)$ and $\epsilon_2 \sim SN(0, 1, 0.5)$			
β_{10}:0	1.7316 (0.1760)	0.4115 (0.6026)	0.2035 (0.4153)
β_{20}:0	0.0162 (−0.0883)	0.0142 (0.0890)	0.0042 (−0.0102)
β_{11}:1	1.9504 (0.1686)	0.0400 (0.0425)	0.0272 (0.0063)
β_{21}:−1	0.0140 (0.0589)	0.0057 (0.0318)	0.0038 (0.0055)
β_{12}:1	1.1483 (0.1871)	0.0426 (0.0430)	0.0303 (0.0114)
β_{22}:−1	0.0157 (0.0639)	0.0062 (0.0364)	0.0039 (0.0076)
w:0.25	0.0053 (−0.0483)	0.0032 (−0.0405)	0.0012 (−0.0037)
Case IV: $\epsilon_1, \epsilon_2 \sim N(0, 1)$ (% 5 outliers)			
β_{10}:0	1.4831 (−0.4157)	7.0063 (2.5017)	1.2509 (0.6932)
β_{20}:0	0.2537 (−0.4938)	0.0412 (−0.1874)	0.1635 (−0.3971)
β_{11}:1	2.9504 (1.4961)	2.4430 (1.4533)	2.5928 (1.4990)
β_{21}:−1	0.0792 (0.2631)	0.0245 (0.1325)	0.0311 (0.1548)
β_{12}:1	2.9498 (1.4958)	2.3507 (1.4210)	2.4870 (1.4631)
β_{22}:−1	0.0760 (0.2565)	0.0222 (0.1247)	0.0286 (0.1467)
w:0.25	0.0099 (−0.0978)	0.0139 (−0.1165)	0.0051 (−0.0681)

Note Value in parentheses indicates the bias

4 Real Data Examples

In this section, we will analyze two real data examples to show the performances of the proposed estimators over the estimators given in literature for the cases with and without outliers.

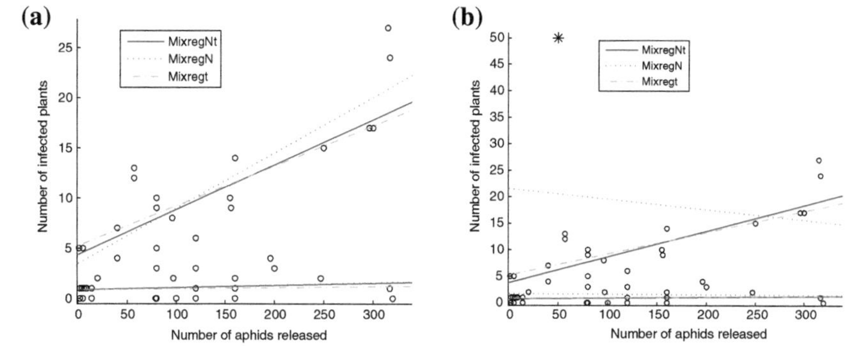

Fig. 2 **a** Fitted mixture regression lines without outlier. **b** Fitted mixture regression lines with outliers at (50, 50)

Table 5 ML estimates and some information criteria for fitting mixture regression models to the aphids data

	MixregN	Mixregt	MixregNt
$\hat{\beta}_{10}$	0.8586	0.8522	0.8648
$\hat{\beta}_{20}$	0.0024	0.0011	0.0022
$\hat{\beta}_{11}$	3.4745	5.2578	4.3813
$\hat{\beta}_{21}$	0.0553	0.0398	0.0448
$\hat{\sigma}_1$	1.1249	0.9127	1.0892
$\hat{\sigma}_2$	3.1153	1.0946	2.2073
$\hat{w}_1$	0.4984	0.5821	0.5128
$\ell(\hat{\boldsymbol{\Theta}})$	**−132.0651**	−137.4716	−133.7615
AIC	**278.1302**	288.9432	281.5230
$CAIC$	**298.6530**	309.4660	302.0458
BIC	**291.6530**	302.4660	295.0458

Note Bold value indicates the smallest values of AIC, CAIC and BIC

Example 1 In this example, we use the aphids data introduced in Sect. 1 which can be accessed by using mixreg package (Turner 2000) in R. We first fit the lines using the estimates based on MixregN, Mixregt and MixregNt. These fitted lines along with the scatter plot of the data are shown in Fig. 2a. We can see that all methods successfully find the groups and give the correct fitted lines. Also, we summarize the ML estimates and the values of some information criteria in Table 5. Note that for the t distribution we assume that $\nu = 2$. We observe that MixregN has the best fit than the other mixture regression models in terms of the Akaike information criterion (AIC) (Akaike 1973), consistent AIC (CAIC) (Bozdogan 1993) and the Bayesian information criterion (BIC) (Schwarz 1978) values.

To see the performances of our estimators when there are outliers in the data, we add five pairs of high leverage outliers at point (50, 50). These points are shown in Fig. 2b by asterisk. Also, the fitted lines and the scatter plot of the data are displayed

Table 6 ML estimates and some information criteria for fitting mixture regression models to the aphids data with five outliers at (50, 50)

	MixregN	Mixregt	MixregNt
$\hat{\beta}_{10}$	1.7334	0.8694	0.7843
$\hat{\beta}_{20}$	−0.0010	0.0013	0.0014
$\hat{\beta}_{11}$	21.5569	5.2484	3.8322
$\hat{\beta}_{21}$	−0.0199	0.0400	0.0484
$\hat{\sigma}_1$	1.6859	1.0449	0.9362
$\hat{\sigma}_2$	16.3307	1.3244	4.0712
$\hat{w}_1$	0.5539	0.5547	0.41212
$\ell(\hat{\boldsymbol{\Theta}})$	−194.0987	−192.1452	**−182.4878**
AIC	402.1974	398.2904	**378.9756**
$CAIC$	416.3749	412.4678	**393.1531**
BIC	423.3749	419.4678	**400.1531**

Note Bold value indicates the smallest values of AIC, CAIC and BIC

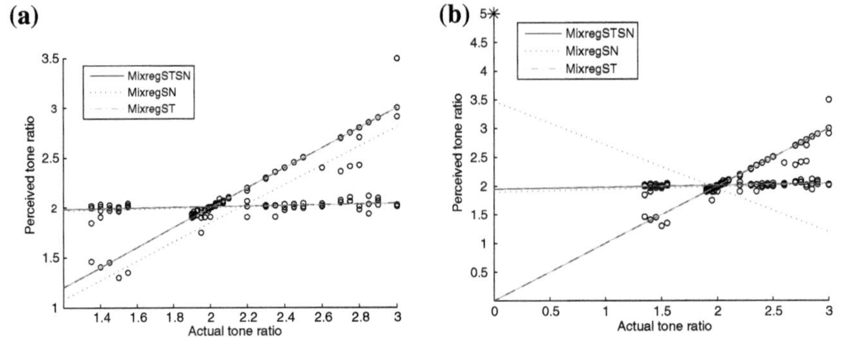

Fig. 3 **a** Fitted mixture regression lines without outlier. **b** Fitted mixture regression lines with outliers at (0, 5)

in Fig. 2b. We give the ML estimates in Table 6. We can see that the fitted lines obtained from MixregN are drastically affected by the outliers. On the other hand, the estimators obtained from Mixregt and MixregNt correctly identifies the groups and fit the regression lines. However, when we compare all methods MixregNt provides the best model in terms of the values of the information criteria.

Example 2. In this example, we use the tone perception data described in Sect. 1 which is given in fpc package (Hennig 2013) in R. This data analyzed by Bai et al. (2012) to model robust mixture regression model. Also, Yao et al. (2014) and Song et al. (2014) used the same data to test performances of the mixture regression estimators based on t and Laplace distributions. The results of these papers show that there should be two groups in the data. We fit mixture of skew normal, mixture of skew t and mixture of skew t and skew normal to check the performances of estimators based on these finite mixture models. We first consider this data without outlier and obtain the

Table 7 ML estimates and some information criteria for fitting mixture regression models to the tone perception data

	MixregSN	MixregST	MixregSTSN
$\hat{\beta}_{10}$	1.9171	1.9491	1.9430
$\hat{\beta}_{20}$	0.0424	0.0318	0.0339
$\hat{\beta}_{11}$	−0.0717	0.0054	0.0030
$\hat{\beta}_{21}$	0.9604	0.9982	0.9988
$\hat{\sigma}_1$	0.0463	0.0393	0.0419
$\hat{\sigma}_2$	0.1883	0.0033	0.0043
$\hat{\lambda}_1$	−0.0100	−0.1666	−0.1692
$\hat{\lambda}_2$	1.7534	0.4465	0.1297
$\hat{w}_1$	0.7006	0.6410	0.6534
$\ell(\hat{\boldsymbol{\Theta}})$	140.5585	211.7766	**215.4246**
AIC	−263.1171	−405.5532	**−412.8491**
$CAIC$	−227.0213	−369.4574	**−376.7534**
BIC	−236.0213	−378.4574	**−385.7534**

Note Bold value indicates the smallest values of AIC, CAIC and BIC

Table 8 ML estimates and some information criteria for fitting mixture regression models to the tone perception data with ten outliers at (0, 5)

	MixregSN	MixregST	MixregSTSN
$\hat{\beta}_{10}$	1.8948	1.9553	1.9450
$\hat{\beta}_{20}$	0.0478	0.0313	0.0339
$\hat{\beta}_{11}$	3.4734	0.0057	0.0031
$\hat{\beta}_{21}$	−0.7579	0.9981	0.9987
$\hat{\sigma}_1$	0.0612	0.0542	0.0562
$\hat{\sigma}_2$	1.2593	0.0031	0.0043
$\hat{\lambda}_1$	−0.2667	−0.2030	−0.1971
$\hat{\lambda}_2$	1.6770	0.4493	0.1297
$\hat{w}_1$	0.7382	0.6759	0.6752
$\ell(\hat{\boldsymbol{\Theta}})$	40.7933	109.3612	**115.0275**
AIC	−63.5867	−200.7225	**−212.0549**
$CAIC$	−26.9101	−164.0459	**−175.3783**
BIC	−35.9101	−173.0459	**−184.3783**

Note Bold value indicates the smallest values of AIC, CAIC and BIC

fitted lines from the mixture models mentioned above. The fitted lines along with the scatter plot are displayed in Fig. 3a. This figure shows that all the models give similar fits. Also, we give the ML estimates and some values of the information criteria in Table 7. The value of the degrees of freedom of the skew t distribution is taken as 2. We see that MixregSTSN gives the best fit than the other mixture regression models in terms of the AIC, CAIC and the BIC values.

To see the performances of the estimators when there are outliers in the data we added ten identical outliers at point (0, 5). The results for the data with outliers are shown in Fig. 3b. Note that the asterisk in this figure shows the location of outliers. It is clear from this figure that the outliers badly affect the estimators obtained from MixregSN. On the other hand, the estimators based on MixregST and MixregSTSN are not affected from the outliers. From the results of information criteria given in Table 8, MixregSTSN has the best fit to model to the tone perception data.

5 Conclusions

In this paper, we have proposed an alternative robust mixture regression model based on the mixture of different type of distributions. We have specifically considered two-component mixture regression based on mixture of t and normal distributions for the symmetric case, and the mixture of skew t and skew normal distributions for the skew case. We have given the EM algorithms for the mixture of different distributions. We have provided a simulation study and two real data examples. The simulation results and the real data examples have shown that the proposed method based on the mixture of different distributions is superior to or comparable with the method based on mixture of the same type of distributions such as mixture of (skew) normal and mixture of (skew) t distribution. If the groups in the data set have different tail behavior using the mixture of different type of distributions should be preferred. For example, in two group case if one of the groups has heavier tails but the other one is not then instead of using mixture of (skew) t distribution one can use mixture of (skew) t and (skew) normal and get the similar result. Using the mixture of t and normal will be computationally less intensive.

Acknowledgments The authors would like to thank referees and the editor for their constructive comments and suggestions that have considerably improved this work. The first author would like to thank the Higher Education Council of Turkey for providing financial support for Ph.D. study in Ankara University. The second author would like to thank the European Commission-JRC and the Indian Statistical Institute for providing financial support to attend the ICORS 2015 in Kolkata in India.

Appendix

To get the conditional expectation of the complete data log-likelihood function given in (8), the following conditional expectations should be calculated given y_j and the current parameter estimate $\hat{\boldsymbol{\Theta}} = (\hat{\boldsymbol{\beta}}_1, \hat{\sigma}_1^2, \hat{\boldsymbol{\beta}}_2, \hat{\sigma}_2^2, \hat{\nu})$

$$\hat{z}_j = E(z_j | y_j, \hat{\boldsymbol{\Theta}}) = \frac{\hat{w}\phi(y_j; \mathbf{x}_j'\hat{\boldsymbol{\beta}}_1, \hat{\sigma}_1^2)}{\hat{w}\phi(y_j; \mathbf{x}_j'\hat{\boldsymbol{\beta}}_1, \hat{\sigma}_1^2) + (1-\hat{w}) f_t(y_j; \mathbf{x}_j'\hat{\boldsymbol{\beta}}_2, \hat{\sigma}_2, \hat{\nu})}, \quad (36)$$

$$\hat{u}_{1j} = E(u_j|y_j, \hat{\boldsymbol{\Theta}}) = \frac{\hat{\nu}+1}{\hat{\nu}+\left(\left(y_j - \mathbf{x}_j^{'}\hat{\boldsymbol{\beta}}_2\right)/\hat{\sigma}_2\right)^2}, \tag{37}$$

$$\hat{u}_{2j} = E(\log u_j|y_j, \hat{\boldsymbol{\Theta}}) = DG\left(\frac{\hat{\nu}+1}{2}\right) - \log\left(\frac{\hat{\nu}}{2} + \frac{\left(y_j - \mathbf{x}_j^{'}\hat{\boldsymbol{\beta}}_2\right)^2}{2\hat{\sigma}_2^2}\right). \tag{38}$$

These conditional expectations will be used in EM algorithm given in Sect. 2.1. Similarly, to obtain the conditional expectation of the complete data log-likelihood function given in (21) the following expectations should be computed given y_j and the current parameter estimate $\hat{\boldsymbol{\Theta}} = (\hat{\boldsymbol{\beta}}_1, \hat{\sigma}_1^2, \hat{\lambda}_1, \hat{\nu}, \hat{\boldsymbol{\beta}}_2, \hat{\sigma}_2^2, \hat{\lambda}_2)$

$$\hat{z}_j = E(z_j|y_j, \hat{\boldsymbol{\Theta}}) = \frac{\hat{w} f_{ST}(y_j; \mathbf{x}_j^{'}\hat{\boldsymbol{\beta}}_1, \hat{\sigma}_1^2, \hat{\lambda}_1, \hat{\nu})}{\hat{w} f_{ST}(y_j; \mathbf{x}_j^{'}\hat{\boldsymbol{\beta}}_1, \hat{\sigma}_1^2, \hat{\lambda}_1, \hat{\nu}) + (1-\hat{w}) f_{SN}(y_j; \mathbf{x}_j^{'}\hat{\boldsymbol{\beta}}_1, \hat{\sigma}_2^2, \hat{\lambda}_2)}, \tag{39}$$

$$\hat{s}_{1j} = E(z_j \tau_j|y_j, \hat{\boldsymbol{\Theta}}) = \hat{z}_j\left(\frac{\hat{\nu}+1}{\hat{\eta}_{1j}^2 + \hat{\nu}}\right)\frac{T_{\hat{\nu}+3}\left(\hat{M}_j\sqrt{\frac{\hat{\nu}+3}{\hat{\nu}+1}}\right)}{T_{\hat{\nu}+1}(\hat{M}_j)}, \tag{40}$$

$$\hat{s}_{2j} = E(z_j \gamma_j \tau_j|y_j, \hat{\boldsymbol{\Theta}}) = \frac{\hat{\delta}_{\lambda_1}(y_j - \mathbf{x}_j^{'}\hat{\boldsymbol{\beta}}_1)\hat{s}_{1j}}{\hat{\sigma}_1} + \frac{\hat{z}_j\sqrt{1-\hat{\delta}_{\lambda_1}^2}}{\pi\hat{\sigma}_1 \hat{f}(y_j)}\left(\frac{\hat{\eta}_{1j}^2}{\hat{\nu}(1-\hat{\delta}_{\lambda_1}^2)} + 1\right)^{-(\frac{\hat{\nu}}{2}+1)}, \tag{41}$$

$$\begin{aligned}\hat{s}_{3j} &= E(z_j \gamma_j^2 \tau_j|y_j, \hat{\boldsymbol{\Theta}}) = \hat{\delta}_{\lambda_1}^2\left(\frac{y_j - \mathbf{x}_j^{'}\hat{\boldsymbol{\beta}}_1}{\hat{\sigma}_1}\right)^2 \hat{s}_{1j} + \hat{z}_j\Bigg\{(1-\hat{\delta}_{\lambda_1}^2) \\ &+ \frac{\hat{\delta}_{\lambda_1}(y_j - \mathbf{x}_j^{'}\hat{\boldsymbol{\beta}}_1)\sqrt{1-\hat{\delta}_{\lambda_1}^2}}{\pi\hat{\sigma}_1^2 \hat{f}(y_j)}\left(\frac{\hat{\eta}_{1j}^2}{\hat{\nu}(1-\hat{\delta}_{\lambda_1}^2)} + 1\right)^{-(\frac{\hat{\nu}}{2}+1)}\Bigg\},\end{aligned} \tag{42}$$

$$\begin{aligned}\hat{s}_{4j} &= E(z_j \log(\tau_j)|y_j, \hat{\boldsymbol{\Theta}}) = \hat{z}_j\Bigg\{DG\left(\frac{\hat{\nu}+1}{2}\right) - \log\left(\frac{\hat{\eta}_{1j}^2 + \hat{\nu}}{2}\right) \\ &+ \left(\frac{\hat{\nu}+1}{\hat{\eta}_{1j}^2 + \hat{\nu}}\right)\left(\frac{T_{\hat{\nu}+3}\left(\hat{\lambda}_1\hat{\eta}_{1j}\sqrt{\frac{\hat{\nu}+3}{\hat{\nu}+\hat{\eta}_{1j}^2}}\right)}{T_{\hat{\nu}+1}\left(\hat{\lambda}_1\hat{\eta}_{1j}\sqrt{\frac{\hat{\nu}+1}{\hat{\nu}+\hat{\eta}_{1j}^2}}\right)} - 1\right) \\ &+ \frac{\hat{\lambda}_1\hat{\eta}_{1j}(\hat{\eta}_{1j}^2 - 1)}{\sqrt{(\hat{\nu}+1)(\hat{\nu}+\hat{\eta}_{1j}^2)^3}}\frac{t_{\hat{\nu}+1}\left(\hat{\lambda}_1\hat{\eta}_{1j}\sqrt{\frac{\hat{\nu}+1}{\hat{\nu}+\hat{\eta}_{1j}^2}}\right)}{T_{\hat{\nu}+1}\left(\hat{\lambda}_1\hat{\eta}_{1j}\sqrt{\frac{\hat{\nu}+1}{\hat{\nu}+\hat{\eta}_{1j}^2}}\right)} \\ &+ \frac{1}{T_{\hat{\nu}+1}\left(\hat{\lambda}_1\hat{\eta}_{1j}\sqrt{\frac{\hat{\nu}+1}{\hat{\nu}+\hat{\eta}_{1j}^2}}\right)}\int_{-\infty}^{\hat{M}_j} g_{\hat{\nu}}(x) t_{\hat{\nu}+1}(x)dx\Bigg\},\end{aligned} \tag{43}$$

$$\hat{t}_{1j} = E(\gamma_j|y_j, \hat{\boldsymbol{\Theta}}) = \hat{\delta}_{\lambda_2}\hat{\eta}_{2j} + \sqrt{1-\hat{\delta}_{\lambda_2}^2}\frac{\phi\left(\hat{\lambda}_2\hat{\eta}_{2j}\right)}{\Phi\left(\hat{\lambda}_2\hat{\eta}_{2j}\right)}, \tag{44}$$

$$\hat{t}_{2j} = E(\gamma_j^2 | y_j, \hat{\boldsymbol{\Theta}}) = 1 - \hat{\delta}_{\lambda_2}^2 + \hat{\delta}_{\lambda_2} \hat{\eta}_{2j} \hat{t}_{1j}, \tag{45}$$

where

$$\hat{\eta}_{1j} = \frac{(y_j - \mathbf{x}_j' \hat{\boldsymbol{\beta}}_1)}{\hat{\sigma}_1}, \hat{\delta}_{\lambda_1} = \frac{\hat{\lambda}_1}{\sqrt{1 + \hat{\lambda}_1^2}},$$

$$\hat{\eta}_{2j} = \frac{(y_j - \mathbf{x}_j' \hat{\boldsymbol{\beta}}_2)}{\hat{\sigma}_2}, \hat{\delta}_{\lambda_2} = \frac{\hat{\lambda}_2}{\sqrt{1 + \hat{\lambda}_2^2}}, \hat{M}_j = \hat{\lambda}_1 \hat{\eta}_{1j} \sqrt{\frac{\hat{\upsilon}}{\hat{\upsilon} + \hat{\eta}_{1j}^2}},$$

$$g_{\hat{\upsilon}}(x) = DG\left(\frac{\hat{\upsilon} + 2}{2}\right) - DG\left(\frac{\hat{\upsilon} + 1}{2}\right) - \log\left(1 + \frac{x^2}{\hat{\upsilon} + 1}\right) + \frac{x^2(\hat{\upsilon} + 1) - \hat{\upsilon} - 1}{(\hat{\upsilon} + 1)(\hat{\upsilon} + 1 + x^2)},$$

$$\hat{f}(y_j) = \hat{w}_1 \frac{2}{\hat{\sigma}_1} t_{\hat{\upsilon}}(\hat{\eta}_{1j}) T_{\hat{\upsilon}+1}(\hat{M}_j) + (1 - \hat{w}_1) \frac{2}{\hat{\sigma}_2} \phi(\hat{\eta}_{2j}) \Phi(\hat{\lambda}_2 \hat{\eta}_{2j}).$$

These conditional expectations will be used in EM algorithm given in Sect. 2.2.

References

Akaike H (1973) Information theory and an extension of the maximum likelihood principle. In: Petrov BN, Caski F (eds) Proceeding of the second international symposium on information theory. Akademiai Kiado, Budapest, pp 267–281

Azzalini A (1985) A class of distributions which includes the normal ones. Scand J Stat 12:171–178

Azzalini A (1986) Further results on a class of distributions which includes the normal ones. Statistica 46:199–208

Azzalini A, Capitaino A (2003) Distributions generated by perturbation of symmetry with emphasis on a multivariate skew t distribution. J R Statist Soc B 65:367–389

Bai X (2010) Robust mixture of regression models. Master's thesis, Kansas State University

Bai X, Yao W, Boyer JE (2012) Robust fitting of mixture regression models. Comput Stat Data An 56:2347–2359

Bozdogan H (1993) Choosing the number of component clusters in the mixture model using a new informational complexity criterion of the inverse-fisher information matrix. Information and classification. Springer, Berlin, pp 40–54

Cohen AC (1984) Some effects of inharmonic partials on interval perception. Music Percept 1:323–349

Dempster AP, Laird NM, Rubin DB (1977) Maximum likelihood from incomplete data via the E-M algorithm. J Roy Stat Soc B Met 39:1–38

Doğru FZ (2015) Robust parameter estimation in mixture regression models. PhD thesis, Ankara University

Doğru FZ, Arslan O (2016) Robust mixture regression based on the skew t distribution. Revista Colombiana de Estadística, accepted

Hennig C (2013) fpc: Flexible procedures for clustering. R Package Version 2.1-5

Henze N (1986) A probabilistic representation of the skew-normal distribution. Scand J Stat 13:271–275

Lin TI, Lee JC, Hsieh WJ (2007) Robust mixture modeling using the skew t distribution. Stat Comput 17:81–92

Liu M, Lin TI (2014) A skew-normal mixture regression model. Educ Psychol Meas 74(1):139–162

Quandt RE (1972) A new approach to estimating switching regressions. J Am Statis Assoc 67:306–310

Quandt RE, Ramsey JB (1978) Estimating mixtures of normal distributions and switching regressions. J Am Stat Assoc 73:730–752

Schwarz G (1978) Estimating the dimension of a model. Ann Stat 6(2):461–464

Song W, Yao W, Xing Y (2014) Robust mixture regression model fitting by laplace distribution. Comput Stat Data An 71:128–137

Turner TR (2000) Estimating the propagation rate of a viral infection of potato plants via mixtures of regressions. J Roy Statist Soc Ser C 49:371–384

Wei Y (2012) Robust mixture regression models using t-distribution. Master's thesis, Kansas State University

Yao W, Wei Y, Yu C (2014) Robust mixture regression using the t-distribution. Comput Stat Data An 71:116–127

Robust Statistical Engineering by Means of Scaled Bregman Distances

Anna-Lena Kißlinger and Wolfgang Stummer

1 Introduction

It is well-known that density-based distances—also known as divergences, disparities, (dis)similarity measures, proximity measures—between two probability distributions serve as useful tools for parameter estimation, testing for goodness-of-fit respectively homogeneity respectively independence, clustering, change point detection, Bayesian decision procedures, as well as for other research fields such as information theory, signal processing including image and speech processing, pattern recognition, feature extraction, machine learning, econometrics, and statistical physics. For some comprehensive surveys on the distance approach to statistics and probability, the reader is referred to the insightful books of, e.g. Liese and Vajda (1987), Read and Cressie (1988), Vajda (1989), Csiszar and Shields (2004), Stummer (2004), Pardo (2006), Liese and Miescke (2008), Basu et al. (2011), Voinov et al. (2013), and the references therein; see also the survey papers of, e.g. Maasoumi (1993), Golan (2003), Liese and Vajda (2006), Vajda and van der Meulen (2010). Distance-based bounds of Bayes risks (e.g. in finance) can be found, e.g. in

Dedicated to the Indian Statistical Institute and to the People of India.

A.-L. Kißlinger
Chair of Statistics and Econometrics, University of Erlangen-Nürnberg,
Lange Gasse 20, 90403 Nürnberg, Germany
e-mail: anna-lena.kisslinger@fau.de

W. Stummer (✉)
Department of Mathematics, University of Erlangen-Nürnberg, Cauerstraße 11,
91058 Erlangen, Germany
e-mail: stummer@math.fau.de

W. Stummer
Affiliated Faculty Member of the School of Business and Economics, University
of Erlangen-Nürnberg, Lange Gasse 20, 90403 Nürnberg, Germany

C. Agostinelli et al. (eds.), *Recent Advances in Robust Statistics: Theory and Applications*, DOI 10.1007/978-81-322-3643-6_5

Stummer and Vajda (2007), see also Stummer and Lao (2012). Amongst others, some important density-based distribution-distance classes are:

- the Csiszar-Ali-Silvey divergences CASD (C. 1963, A.-S. 1966): this includes, e.g. the total variation distance, exponentials of Renyi cross-entropies (Hellinger integrals), and the power divergences (also known as $\alpha-$entropies, Cressie-Read measures, Tsallis cross-entropies); the latter cover, e.g. the Kullback–Leibler information divergence (relative entropy), the (squared) Hellinger distance, the Pearson chi-square divergence;
- the "classical" Bregman distances CBD (see, e.g. Bregman (1967), Csiszar (1991), Csiszar (1994), Csiszar (1995), Pardo and Vajda (1997), Pardo and Vajda (2003)): this includes, e.g. the density power divergences DPD (also known as Basu-Harris-Hjort-Jones BHHJ distances, cf. Basu et al. (1998)) with the squared L_2-norm as special case.

The modern use of density-based distances between distributions with a view towards robustness investigations started with the seminal paper of Beran (1977), which was considerably extended by Lindsay (1994), Basu and Lindsay (1994); the latter two articles develop insightful comparisons between the robustness performance of some (in effect) CASD distances, in terms of the important concept of the *residual adjustment function* RAF (having the Pearson residual between the data and the candidate model as its argument). In contrast, the robustness properties of the above-mentioned DPDs were studied by Basu et al. (1998). The growing literature of these two research lines is comprehensively summarized in the book of Basu et al. (2011); more recently, Basu et al. (2013, 2015a) develop the asymptotic distribution of DPD test statistics, Ghosh and Basu (2013, 2014) apply the DPD family for robust and efficient parameter estimation for linear regressions and for censored data, whereas Basu et al. (2015b) use the DPD for the development of robust tests for the equality of two normal means. Concerning some recent progress of divergences, Stummer (2007) as well as Stummer and Vajda (2012) introduced the concept of *scaled Bregman divergences/distances* SBD, which enlarges all the above-mentioned (nearly disjoint) CASD and CBD divergence classes at once. Hence, the SBD class constitutes a broad framework for dealing with a wide range of data analyses in a well-structured way; for each concrete data analysis, the free choice of the two major SBD building blocks (generator, scaling measure) implies much flexibility for interdisciplinary situation-based inference, see e.g. Kißlinger and Stummer (2013, 2015a, b) for corresponding exemplary contexts of applicability for parameter estimation, goodness-of-fit testing, model search, change detection, etc. Combining the insights of the preceding explanations suggests that all the above-mentioned robustness and efficiency investigations can be covered at once and further extended by means of SBDs. Accordingly, the main goals of this paper are (i) to build up a SBD-based bivariate statistical-engineering paradigm for the goal-oriented design of new outlier- and inlier-robust statistical inference tools, and (ii) to derive asymptotic distributions of the corresponding SBD test statistics. To achieve this, we first develop in Sect. 2 some new classes of *scale connectors*. Subsequently, in order to obtain a comfortable interpretability of the underlying robustness structure—through 3D visualization—we introduce in Sect. 3

the concept of *density-pair adjustment functions* DAF(0) and DAF(1) of zeroth and first order; the latter is a flexible bivariate extension of the above-mentioned (univariate) RAF. Finally, for the finite discrete case the asymptoticity (ii) is investigated in Sect. 4. Throughout this paper, we present numerous 3D plots in order to illustrate the immediate applicability of our methods.

2 Scaled Bregman Divergences

Suppose that we want to measure the divergence (distance, dissimilarity, proximity) $D(P, Q)$ between two probability distributions P, Q on a space $\mathscr{X}$ (with $|\mathscr{X}| \geq 2$) equipped with a $\sigma-$algebra $\mathscr{A}$. Typically, one has the following constellations:

- P and Q are both "theoretical distributions", e.g. $P = Poisson(a_1)$ and $Q = Poisson(a_2)$;
- P and Q are both (random sample versions of) "data-derived distributions", which appear, e.g. in the context of change detection, two-sample testing; for instance, $P := P_N^{emp} := \frac{1}{N} \cdot \sum_{i=1}^{N} \delta_{X_i}[\cdot]$ is the histogram-conform empirical distribution[1] of an iid sample $X_1, \ldots, X_N$ of size N from P_{θ_1} and $Q := P_M^{emp}$ is the empirical distribution of an iid sample $\widetilde{Y}_1, \ldots, \widetilde{Y}_M$ from P_{θ_2};
- P is a data-derived distribution and Q is a theoretical distribution; e.g. in the context of minimum distance estimation or goodness-of-fit testing, $P = P_N^{emp}$ and $Q = P_\theta$ is a hypothetical candidate for the unknown underlying true sampling distribution P_{θ_1}. Thus, $D(P_N^{emp}, P_\theta)$ measures the discrepancy between the "pattern" of the observed data and the "pattern" predicted by the candidate model.

Since the ultimate purposes of a (divergence-based) statistical inference may vary from case to case, some goal-oriented situation-based well-structured flexibility can be obtained by using a toolbox $\mathscr{D} := \{D_{\phi,M}(P, Q) : \phi \in \Phi, M \in \mathscr{M}\}$ of divergences which is far-reaching due to various different choices of a "generator" $\phi \in \Phi$ and a "scaling measure" $M \in \mathscr{M}$. In particular, this should also cover robustness issues (in a wide sense). To find good choices of $\mathscr{D}$ is one of the purposes of "statistical engineering". One possible candidate for a (density-based) wide divergence family $\mathscr{D}$ is the concept of scaled Bregman divergences SBD of Stummer (2007), Stummer and Vajda (2012). To describe this, for the sake of brevity we deal with the generator class $\Phi = \Phi_{C_1}$ of functions $\phi : (0, \infty) \mapsto \mathbb{R}$ which are continuously differentiable with derivative ϕ', strictly convex and continuously extended to $t = 0$, and (w.l.o.g.) satisfy $\phi(1) = 0$, $\phi'(1) = 0$. Furthermore, for a fixed σ-finite ("reference") measure λ on $\mathscr{X}$ we denote by $\mathscr{M}_\lambda$ resp. $\mathscr{P}_\lambda$ the family of all $\sigma-$finite measures M on $\mathscr{X}$ resp. all probability measures (distributions) P on $\mathscr{X}$ having densities $m = \frac{dM}{d\lambda} \geq 0$ resp. $p = \frac{dP}{d\lambda} \geq 0$ with respect to λ. For $Q \in \mathscr{P}_\lambda$, we write $q = \frac{dQ}{d\lambda} \geq 0$. Within such a context (and even for non-differentiable generators ϕ), Stummer (2007), Stummer and Vajda (2012) introduced the following framework of statistical distances:

[1]Notice that δ_y is Dirac's one-point distribution at y (i.e. $\delta_y[A] = 1$ iff $y \in A$ and $\delta_y[A] = 0$ else).

Definition 1 Let $\phi \in \Phi_{C_1}$ and λ be σ-finite measure on $\mathscr{X}$. Then the Bregman divergence of $P, Q \in \mathscr{P}_\lambda$ scaled by $M \in \mathscr{M}_\lambda$ is defined by

$$0 \leq D_{\phi,M}(P,Q) := B_\phi\left(P,Q \,|\, M\right)$$

$$:= \int_{\mathscr{X}} \left[\phi\left(\frac{p(x)}{m(x)}\right) - \phi\left(\frac{q(x)}{m(x)}\right) - \phi'\left(\frac{q(x)}{m(x)}\right)\cdot\left(\frac{p(x)}{m(x)} - \frac{q(x)}{m(x)}\right)\right] \mathrm{d}M(x) \qquad (1)$$

$$= \int_{\mathscr{X}} \left[m(x)\cdot\left\{\phi\left(\frac{p(x)}{m(x)}\right) - \phi\left(\frac{q(x)}{m(x)}\right)\right\} - \phi'\left(\frac{q(x)}{m(x)}\right)\cdot(p(x)-q(x))\right] \mathrm{d}\lambda(x). \quad (2)$$

To guarantee the existence of the integrals in (1), (2) (with possibly infinite values), the zeros of p, q, m have to be combined by proper conventions (see, e.g. Kißlinger and Stummer (2013) for some discussions; the full details will appear elsewhere).

Notice that—for fixed M—one gets $B_\phi\left(P,Q \,|\, M\right) = B_{\tilde{\phi}}\left(P,Q \,|\, M\right)$ for any $\tilde{\phi}(t) := \phi(t) + c_1 + c_2 \cdot t \quad (t \in]0,\infty[)$ with constants $c_1, c_2 \in \mathbb{R}$. Moreover, there exist "essentially different" pairs (ϕ, M) and $(\breve{\phi}, \breve{M})$ (where $\phi(t) - \breve{\phi}(t)$ is nonlinear in t) for which $B_\phi\left(P,Q \,|\, M\right) = B_{\breve{\phi}}\left(P,Q \,|\, \breve{M}\right)$, c.f. Remark 1a below. For power functions

$$\phi(t) := \phi_\alpha(t) := \frac{t^\alpha - 1}{\alpha(\alpha-1)} - \frac{t-1}{\alpha-1} \geq 0\,, \qquad t \in]0,1[,\ \alpha \in \mathbb{R}\backslash\{0,1\}\,, \qquad (3)$$

one obtains from (2) the scaled Bregman power distances

$$B_{\phi_\alpha}\left(P,Q \,|\, M\right) = \int_{\mathscr{X}} \frac{m(x)^{1-\alpha}}{\alpha-1}\cdot\left(\frac{p(x)^\alpha}{\alpha} + \frac{(\alpha-1)\cdot q(x)^\alpha}{\alpha} - p(x)\cdot q(x)^{\alpha-1}\right)\mathrm{d}\lambda(x) \qquad (4)$$

(cf. Stummer and Vajda (2012), Kißlinger and Stummer (2013)), especially

$$B_{\phi_2}\left(P,Q \,|\, M\right) = \frac{1}{2}\int_{\mathscr{X}} \frac{(p(x)-q(x))^2}{m(x)}\,\mathrm{d}\lambda(x)\,.$$

In the discrete case where $\mathscr{X} = \{x_1, x_2, \ldots\}$ is finite or countable and $\lambda := \lambda_{count}$ is the counting measure (i.e., $\lambda_{count}[\{x_k\}] = 1$ for all k), then p, q, m are (probability) mass functions and (2) becomes

$$0 \leq B_\phi\left(P,Q \,|\, M\right) \qquad (5)$$

$$= \sum_{x\in\mathscr{X}} \left[m(x)\cdot\left\{\phi\left(\frac{p(x)}{m(x)}\right) - \phi\left(\frac{q(x)}{m(x)}\right)\right\} - \phi'\left(\frac{q(x)}{m(x)}\right)\cdot(p(x)-q(x))\right].$$

For instance, if $P := P_N^{emp} := \frac{1}{N}\cdot\sum_{i=1}^N \delta_{X_i}[\cdot]$ is the above-mentioned data-derived empirical distribution of an iid sample $X_1, \ldots, X_N$ of size N, the corresponding probability mass functions are the relative frequencies $p(x) = p_N^{emp}(x) = \frac{1}{N}\cdot\#\{i \in \{1,\ldots,N\} : X_i = x\}$; if $Q = P_\theta$ is a hypothetical candidate model, then $q(x) =$

$p_\theta(x)$. In contrast, for $\mathscr{X} = \mathbb{R}$ and Lebesgue measure $\lambda := \lambda_{Leb}$, one gets p, q, m as "classical" (probability) densities and (2) reads as "classical" Lebesgue (Riemann) integral. As an example, take the standard Gaussian $p(x) = \frac{1}{2\pi} exp\{-x^2/2\}$.

Returning back to the general context, for applicability purposes we aim here for a wide class $\mathscr{M}$ of scaling measures M such that the outcoming SBDs $B_\phi\,(P, Q \,|\, M)$

- *cover much more* than the frequently used concepts of Csiszar-Ali-Silvey divergences CASD and classical Bregman distances CBD,
- can be used to tackle robustness (and many other) issues in a well-structured, finely nuanced way,
- lead (despite of the involved generality) to concrete explicit asymptotic results for data analyses where the sample size grows to infinity.

Correspondingly, in the following we confine ourselves to the special subframework $B_\phi\,(P, Q \,|\, W(P, Q))$ with scaling measures of the form $M = W(P, Q)$ in the sense that $m(x) = w(p(x), q(x)) \geq 0$ ($\lambda-$a.a. $x \in \mathscr{X}$) for some (measurable) "scale-connector" $w : [0, \infty[\times[0, \infty[\mapsto [0, \infty]$ between the densities $p(x)$ and $q(x)$ (where w is strictly positive on $]0, \infty[\times]0, \infty[$). In the discrete case we shall restrict w to $[0, 1] \times [0, 1]$ (take, e.g. $P := P_N^{emp}$ and $Q = P_\theta$ as a running example); accordingly, the underlying construction principle and 3D plotting technique of $B_{\phi_\alpha}\big(P, Q \,|\, W_\beta(P, Q)\big)$ is illustrated in Fig. 1, where we allow w to depend on a parameter β. Within such a context, the following special choices of scale connectors $w(\cdot, \cdot)$ are of particular interest:

1. no scaling: $M = 1$, i.e. $w(u, v) := w_{no}(u, v) = 1$ for all $u, v \in [0, \infty[$. Then,

$$B_\phi\,(P, Q \,|\, 1) = \int_{\mathscr{X}} \big[\phi(p(x)) - \phi(q(x)) - \phi'(q(x)) \cdot (p(x) - q(x))\big]\, \mathrm{d}\lambda(x) \quad (6)$$

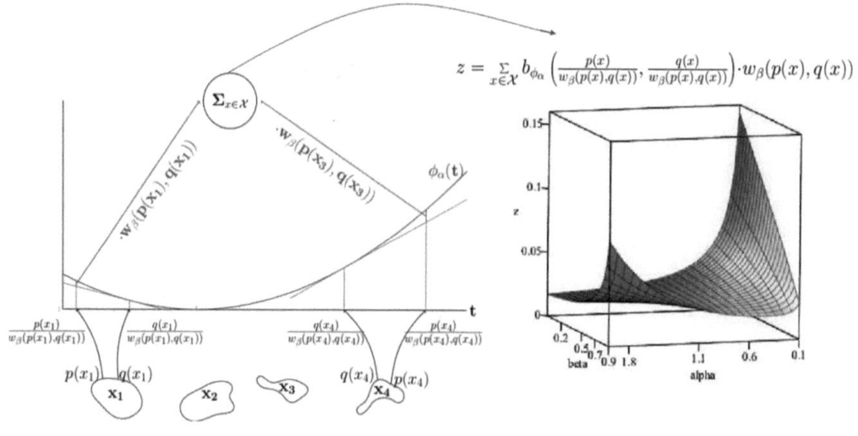

Fig. 1 Construction of scaled Bregman distance family

is the classical Bregman distance CBD between P and Q, generated by ϕ. For the particular choice $\phi = \phi_\alpha$ (cf. (3)), $B_{\phi_\alpha}(P, Q \mid 1)$ (cf. (4)) is a multiple of the $\alpha-$order density power divergence DPD_α (also known as BHHJ divergence) of Basu et al. (1998); see also Basu et al. (2011, 2013, 2015a, b), Ghosh and Basu (2013, 2014), for recent applications. Notice that, e.g. for finite space $\mathscr{X}$, $B_{\phi_2}(P, Q \mid 1)$ corresponds to the squared L_2-norm.

2. multiple idempotency scaling: the scale connector $w(\cdot, \cdot)$ is arbitrary with the only constraint that

$$\exists c > 0 \; \forall v \in [0, \infty[\quad w(v, v) = c \cdot v \,. \tag{7}$$

For instance, this will turn out to be important for obtaining a "straight" (i.e., unmixtured) chi-square distribution for the asymptotics of corresponding scaled Bregman distances in an i.i.d. sample context, see Subcase 3 in Sect. 4 below. Notice that for two multiple idempotency scalings w_1 and w_2, also $c_1 w_1 + c_2 w_2$ ($c_1, c_2 \geq 0$ with $c_1 + c_2 > 0$), $\min\{w_1, w_2\}$ and $\max\{w_1, w_2\}$ are multiple idempotency scalings; however, the latter two may not inherit the differentiability properties of w_1 and w_2 which may lead to complications in asymptoticity assertions. Let us mention some important special cases (see the figures in Fig. 2 for illustration in the discrete case where $u, v \in [0, 1]$):

2a. Scalings of the form

$$w(u, v) := w_{\beta, f}(u, v) := c \cdot f^{-1}\left(\beta \cdot f(u) + (1 - \beta) \cdot f(v)\right) \,, u \in [0, \infty[\,, v \in [0, \infty[\tag{8}$$

for some strictly monotone function $f : [0, \infty[\to [0, \infty[$, $\beta \in [0, 1]$. This means that the scale connector $w_{\beta, f}$ is a positive multiple of a weighted generalized quasi-linear mean WGQLM (between u and v); for a general comprehensive study on WGQLMs, see e.g. Grabisch et al. (2009). For fixed $\beta \in [0, 1]$ and $u, v \in [0, \infty[$, we derive from (8) the following useful subsubcases:

2ai. multiple of weighted rth-power mean WRPM, $r \in \mathbb{R} \backslash \{0\}$:

$$w_{\beta, f_r}(u, v) := w_{\beta, r}(u, v) := c \cdot \left(\beta \cdot u^r + (1 - \beta) \cdot v^r\right)^{\frac{1}{r}} \,, \quad \text{where } f_r(z) := z^r \,.$$

Notice that there is a one-to-one correspondence between $w_{\beta, r}(u, v)$ and $w_{\beta_1, \beta_2, r}(u, v) := (\beta_1 \cdot u^r + \beta_2 \cdot v^r)^{1/r}$ with $\beta_1, \beta_2 \geq 0$ such that $\beta_1 + \beta_2 > 0$. For the rest of this paragraph 2ai, we assume $c = 1$. The case $r = 1$ corresponds to weighted-arithmetic-mean scaling (mixture scaling) $w_{\beta, 1}(u, v) = \beta \cdot u + (1 - \beta) \cdot v$ (cf. Fig. 2b). In particular, with $w_{0,1}(u, v) = v$ (cf. Fig. 2a) one obtains the corresponding Csiszar-Ali-Silvey ϕ divergence CASD

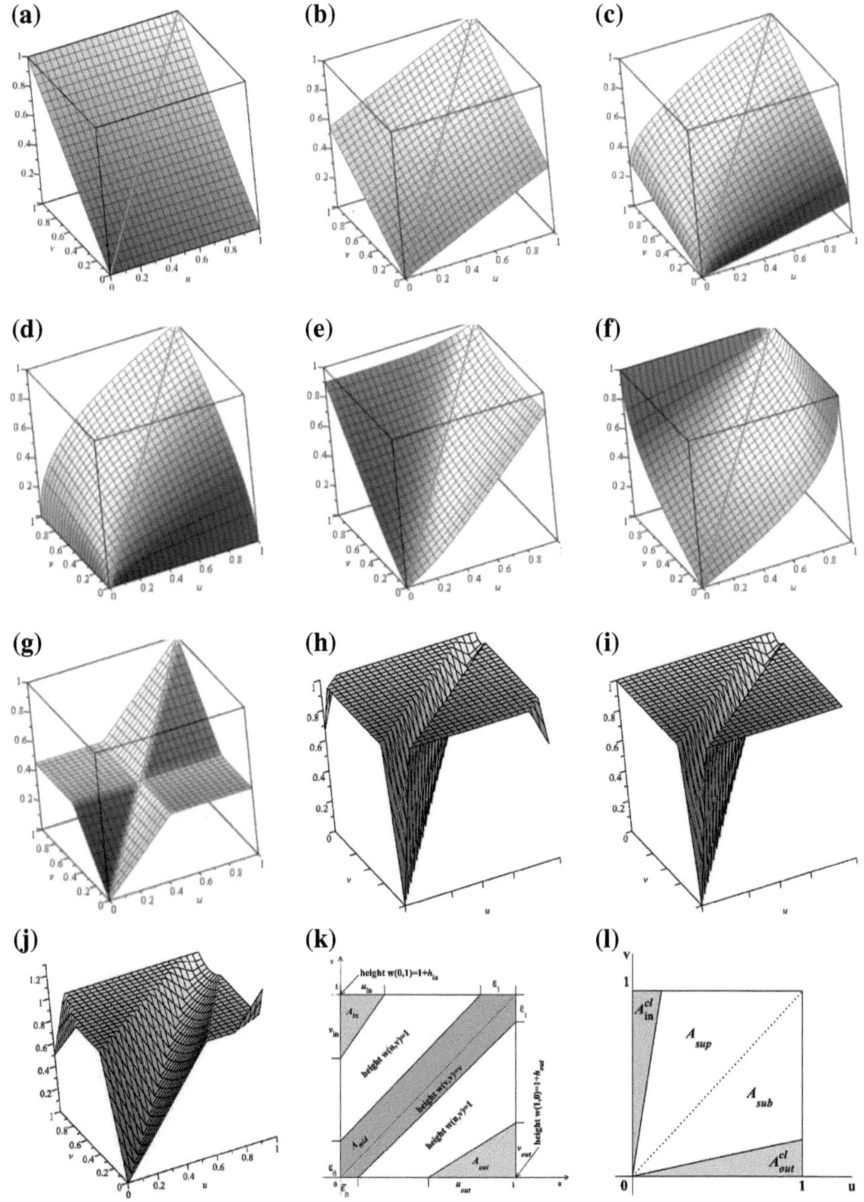

Fig. 2 Some scale connectors; $c = 1$. **a** $w_{0,1}(u, v) = v$ (all Csiszar-Ali-Silvey divergences). **b** $w_{0.45,1}(u, v) = 0.45 \cdot u + 0.55 \cdot v$. **c** $w_{0.45,0.5}(u, v) = (0.45\sqrt{u} + 0.55\sqrt{v})^2$. **d** $w_{0.45,0}(u, v) = u^{0.45} \cdot v^{0.55}$. **e** $w_{0.45,\tilde{f}_6}(u, v) = \frac{1}{6} \log\left(0.45e^{6u} + 0.55e^{6v}\right)$. **f** $w_E(u, v) = \frac{1 + uv - \sqrt{(1 - u^2)(1 - v^2)}}{u + v}$. **g** $w_{0.45}^{med}(u, v) = \text{med}\{\min\{u, v\}, 0.45, \max\{u, v\}\}$. **h** $w_{adj}(u, v)$ with $h_{in} = -0.33$, $h_{out} = -0.25$, etc. **i** $w_{adj}(u, v)$ with $h_{in} = 0$, $h_{out} = 0$, etc. **j** $w_{adj}^{smooth}(u, v)$ with $h_{in} = -0.5$, $h_{out} = 0.3$, $\delta = 10^{-7}$, etc. **k** Parameter description for $w_{adj}(u, v)$, cf. (**h**), (**i**), (**j**). **l** classical outlier/inlier areas A_{out}^{cl} resp. A_{in}^{cl}

$$B_{\phi}\left(P, Q \mid W_{0,1}(P, Q)\right) = B_{\phi}\left(P, Q \mid Q\right) = \int_{\mathscr{X}} q(x)\, \phi\left(\frac{p(x)}{q(x)}\right) \mathrm{d}\lambda(x) \tag{9}$$

(cf. Stummer (2007), Stummer and Vajda (2012)). Hence, in our context *every* CASD appears as a subsubcase. For $\phi(t) := \phi_1(t) := t \log t + 1 - t$ one arrives at the Kullback–Leibler information divergence KL (relative entropy)

$$B_{\phi_1}\left(P, Q \mid W_{0,1}(P, Q)\right) = B_{\phi_1}\left(P, Q \mid Q\right) = \int_{\mathscr{X}} p(x)\, \log\left(\frac{p(x)}{q(x)}\right) \mathrm{d}\lambda(x)\,.$$

Moreover, the choice $\phi(t) := \phi_{1/2}(t)$ (cf. (3)) leads to the (squared) Hellinger distance $B_{\phi_{1/2}}\left(P, Q \mid W_{0,1}(P, Q)\right) = 2\int_{\mathscr{X}} (\sqrt{p(x)} - \sqrt{q(x)})^2\, \mathrm{d}\lambda(x)$, whereas for $\phi(t) := \phi_2(t)$ (cf. (3)) we end up with Pearson's chi-square divergence

$$B_{\phi_2}\left(P, Q \mid W_{0,1}(P, Q)\right) = B_{\phi_2}\left(P, Q \mid Q\right) = \frac{1}{2}\int_{\mathscr{X}} \frac{(p(x) - q(x))^2}{q(x)}\, \mathrm{d}\lambda(x)\,.$$

For general $\beta \in [0, 1]$, we deduce

$$B_{\phi_2}\left(P, Q \mid W_{\beta,1}(P, Q)\right) = \frac{1}{2}\int_{\mathscr{X}} \frac{(p(x) - q(x))^2}{\beta p(x) + (1-\beta) q(x)}\, \mathrm{d}\lambda(x)$$

which is (the general-space-form of) the blended weight chi-square divergence BWCD of Lindsay (1994); in particular, $B_{\phi_2}\left(P, Q \mid W_{1,1}(P, Q)\right)$ is Neyman's chi-square divergence.
The more general divergences $B_{\phi_\alpha}\left(P, Q \mid W_{\beta,1}(P, Q)\right)$ were used in Kißlinger and Stummer (2013). As far as r is concerned, other interesting special cases (in addition to $r = 1$) of $B_{\phi}\left(P, Q \mid W_{\beta,r}(P, Q)\right)$ are the scaling by (multiple of) weighted quadratic mean ($r = 2$), weighted harmonic mean ($r = -1$), and weighted square root mean ($r = 1/2$, cf. Fig. 2c). For instance, the divergence

$$B_{\phi_2}\left(P, Q \mid W_{\beta,1/2}(P, Q)\right) = \frac{1}{2}\int_{\mathscr{X}} \frac{(p(x) - q(x))^2}{\left(\beta\sqrt{p(x)} + (1-\beta)\sqrt{q(x)}\right)^2}\, \mathrm{d}\lambda(x)$$

corresponds to (the general-space-form of) the blended weight Hellinger distance of Lindsay (1994), Basu and Lindsay (1994).

2aii. multiple of weighted geometric mean WGM:

$$w_{\beta,f_0}(u, v) := w_{\beta,0}(u, v) := c \cdot u^{\beta} \cdot v^{1-\beta}, \quad \text{where } f_0(z) := \log(z)$$

(cf. Fig. 2d). Notice that $\lim_{r\to 0} w_{\beta,r}(u,v) = w_{\beta,0}(u,v)$. For an exemplary application of $B_{\phi_\alpha}\left(P, Q \mid W_{\beta,0}(P,Q)\right)$ to model search for (auto) regressions, see Kißlinger and Stummer (2015b).

2aiii. multiple of weighted exponential mean WEM, $r \in \mathbb{R}\backslash\{0\}$ (cf. Fig. 2e):

$$w_{\beta,\tilde{f}_r}(u,v) := \frac{c}{r} \cdot \log\left(\beta e^{ru} + (1-\beta)e^{rv}\right), \quad \text{where } \tilde{f}_r(z) := \exp(rz). \tag{10}$$

2aiv. multiple of transformed Einstein sum: if u, v are restricted to [0,1]—as it is the case of probability mass functions in (5)—then for $u + v > 0$

$$w_E(u,v) = c \cdot \frac{1 + uv - \sqrt{(1-u^2)(1-v^2)}}{u+v} \quad \text{where } f(z) := \log\left(\frac{1+z}{1-z}\right)$$

(cf. Fig. 2f). Within a context not concerned with probability distances, the special case $c = 1$ can be found, e.g. in Grabisch et al. (2009).

2b. (further) limits of weighted rth-power means: for any $\beta \in [0, 1]$

$$w_{\beta,\infty}(u,v) := \lim_{r\to\infty} w_{\beta,r}(u,v) = c \cdot \max\{u,v\}$$
$$w_{\beta,-\infty}(u,v) := \lim_{r\to-\infty} w_{\beta,r}(u,v) = c \cdot \min\{u,v\}\,.$$

Notice the following bounds:

$$\forall c \in]0,\infty[\ \forall \beta \in [0,1]:\ c \cdot \min\{u,v\} \le w_{\beta,r}(u,v) \le c \cdot \max\{u,v\}\,, \quad u,v \in [0,\infty[.$$

2c. $\beta-$median BM, $\beta \in [0,\infty[$ (cf. Fig. 2g):

$$w_\beta^{med} := c \cdot \text{med}\{\min\{u,v\}, \beta, \max\{u,v\}\} \tag{11}$$

where $med(x_1, x_2, x_3)$ denotes the second smallest of the three numbers x_1, x_2, x_3.

2d. flexible robustness-adjustable scale connector ("robustness adjuster"): here, we confine ourselves to $u, v \in [0, 1]$ which holds for the discrete case where $u = p(x)$, $v = q(x)$ are probability masses (cf. (5)). For parameters $\underline{\varepsilon}_0$, $\underline{\varepsilon}_1$, $\overline{\varepsilon}_0$, $\overline{\varepsilon}_1$, u_{in}, v_{in}, u_{out}, v_{out}, h_{in}, h_{out} satisfying the constraints $0 \le u_{in} \le 1 - \underline{\varepsilon}_1 \le 1$, $0 \le \overline{\varepsilon}_0 \le 1 - u_{out} \le 1$, $0 \le \underline{\varepsilon}_0 \le 1 - v_{in} \le 1$, $0 \le v_{out} \le 1 - \overline{\varepsilon}_1 \le 1$, $h_{in} > -1$, $h_{out} > -1$, we define $\overline{\varepsilon}_v := \overline{\varepsilon}_0 + v(\overline{\varepsilon}_1 - \overline{\varepsilon}_0)$, $\underline{\varepsilon}_v := \underline{\varepsilon}_0 + v(\underline{\varepsilon}_1 - \underline{\varepsilon}_0)$ and

$$\begin{aligned} w(u,v) := w_{adj}(u,v) \ &:= \ w_{mid}(u,v) + w_{in}(u,v) + w_{out}(u,v) \\ &:= 1 + (1-v) \cdot \Bigg\{ |u-v| \cdot \left(\frac{1}{\overline{\varepsilon}_v} \mathbb{1}_{[0,\overline{\varepsilon}_v]}(u-v) + \frac{1}{\underline{\varepsilon}_v} \mathbb{1}_{[-\underline{\varepsilon}_v,0)}(u-v) \right) \\ &\quad - \mathbb{1}_{[-\underline{\varepsilon}_v;\overline{\varepsilon}_v]}(u-v) \Bigg\} \end{aligned}$$

$$+sgn(h_{in}) \cdot \max\left(|h_{in}|\left(1 - \frac{1}{v_{in}} - \frac{u}{u_{in}} + \frac{v}{v_{in}}\right), 0\right)$$
$$+sgn(h_{out}) \cdot \max\left(|h_{out}|\left(1 - \frac{1}{u_{out}} + \frac{u}{u_{out}} - \frac{v}{v_{out}}\right), 0\right), \quad (12)$$

cf. Fig. 2k. Notice that $w(\cdot,\cdot)$ takes the form of the sum of a "plateau $w_{mid}(u,v)$ of height 1 with increasing rift valley around the diagonal (v,v)", a "pyramid $w_{in}(u,v)$ of (possibly negative) height h_{in} around $(u,v) = (0,1)$" and a "pyramid $w_{out}(u,v)$ of (possibly negative) height h_{out} around $(u,v) = (1,0)$"; depending on the values of the parameters, this may look like an edged version of a butterfly, starship or sailplane; see, e.g. Fig. 2h (where here and for (i), (j) we have fixed the parameters $\overline{\varepsilon}_0 = 0.12$, $\overline{\varepsilon}_1 = 0.15$, $\underline{\varepsilon}_0 = 0.13$, $\underline{\varepsilon}_1 = 0.1$, $v_{in} = v_{out} = u_{in} = u_{out} = 0.1$). Furthermore, $w_{adj}(v,v) = v$ for all $v \in [0,1]$. If $h_{in} = h_{out} = 0$ (see Fig. 2i) and additionally $\underline{\varepsilon}_0 = \underline{\varepsilon}_1 = \overline{\varepsilon}_0 = \overline{\varepsilon}_1 =: \varepsilon$ with extremely small $\varepsilon > 0$ (e.g. less than the rounding-errors-concerning machine epsilon of your computer), then for all practical purposes $w_{adj}(\cdot,\cdot)$ is "equal" to the no-scaling connector $w_{no}(u,v)$ and thus $B_\phi\left(P, Q \mid W_{adj}(P,Q)\right)$ is "computationally indistinguishable" from the unscaled classical Bregman divergence $B_\phi(P, Q \mid 1)$ (e.g. for $\phi = \phi_\alpha$, the latter are the DPD$_\alpha$). Of course, the scale connector $w_{adj}(\cdot,\cdot)$ is generally non-smooth which may be uncomfortable for asymptotics properties (see Sect. 4 below). However, one can generate a smoothed version $w_{adj}^{smooth}(u,v) = \int_{\mathbb{R}^2} w_{adj}(\xi_1,\xi_2)\; g_\delta(u-\xi_1, v-\xi_2)\, \mathrm{d}(\xi_1,\xi_2)$ in terms of a mollifier $g_\delta(\cdot,\cdot)$ with some small tuning parameter $\delta > 0$, e.g. $g_\delta(z_1,z_2) := \frac{c}{\delta^2}\exp\{-1/(1-(z_1^2+z_2^2)/\delta^2)\}\mathbb{1}_{[0,\delta^2[}(z_1^2+z_2^2)$ where $c > 0$ is the normalizing constant; see Fig. 2j where we have used $\delta = 10^{-7}$, together with parameters $\overline{\varepsilon}_0 = 0.6$, $\overline{\varepsilon}_1 = 0.2$, $\underline{\varepsilon}_0 = 0.3$, $\underline{\varepsilon}_1 = 0.1$, $v_{in} = 0.4$, $v_{out} = u_{in} = 0.1$, $u_{out} = 0.2$. For "practical purposes", $w_{adj}^{smooth}(\cdot,\cdot)$ "coincides" with $w_{adj}(\cdot,\cdot)$. An analogous smoothing can be done for the scale connectors in 2b and 2c.

2e. Following the lines of Grabisch et al. (2009) on general aggregation functions, one can construct scale connectors which satisfy (7) by means of $w(u,v) = c \cdot \Upsilon_H^{-1}(H(u,v))$ for any (measurable) function $H : [0,\infty[\times[0,\infty[\mapsto [0,\infty]$ for which $z \mapsto \Upsilon_H(z) := H(z,z)$ is strictly increasing. For scale connectors where u, v are restricted to [0,1], this applies analogously.

Remark 1 (a) Notice that the scale connectors w in 2ai, 2aii, 2b can be written in the form $w(u,v) = c \cdot v \cdot h\left(\frac{u}{v}\right)$ for some function $h : [0,\infty] \mapsto [0,\infty]$ (i.e. the plot of $(u,v) \mapsto w(u,v)/v$ is constant along every line through the origin) and hence the corresponding scaled Bregman divergences satisfy $B_{\phi_\alpha}(P, Q \mid W(P,Q)) = B_{\widetilde{\phi}_\alpha}(P, Q \mid Q)$ $(\alpha \in \mathbb{R}\backslash\{0,1\})$, i.e. they can be interpreted as a CASD with non-obvious generator $\widetilde{\phi}_\alpha(t) := \frac{1}{\alpha-1}\left[h(t)^{1-\alpha} \cdot \left\{\frac{t^\alpha}{\alpha} - t + \frac{\alpha-1}{\alpha}\right\}\right]$. For instance, the BWCD is representable as $B_{\phi_2}\left(P, Q \mid W_{\beta,1}(P,Q)\right) = B_{\widetilde{\phi}_2}(P, Q \mid Q)$ where $\widetilde{\phi}_2(t) = (1-t)^2/(2\cdot((1-\beta)+\beta\cdot t))$ turns out to be Rukhin's generator (cf. Rukhin (1994), see, e.g. also Marhuenda et al. (2005), Pardo (2006)). Moreover, for the (squared)

Hellinger distance one has $B_{\phi_2}\left(P, Q \mid W_{1/2,1/2}(P, Q)\right) = B_{\phi_{1/2}}\left(P, Q \mid W_{0,1}(P, Q)\right)$.

(b) The scale connectors $w(u, v)$ in 2aiii, 2aiv, 2c, 2d can NOT be written in the form of $c \cdot v \cdot h\left(\frac{u}{v}\right)$, and hence the CASD-connection of Remark 1a does not apply.

(c) In case of symmetric scale connectors $w(u, v) = w(v, u)$, one can produce symmetric divergences by means of either

$$\begin{aligned}
&B_\phi\left(P, Q \mid W(P, Q)\right) + B_\phi\left(Q, P \mid W(P, Q)\right),\\
&\max\{B_\phi\left(P, Q \mid W(P, Q)\right), B_\phi\left(Q, P \mid W(P, Q)\right)\},\\
&\min\{B_\phi\left(P, Q \mid W(P, Q)\right), B_\phi\left(Q, P \mid W(P, Q)\right)\};
\end{aligned}$$

this also works for $\phi(t) = \phi_1(t)$ together with arbitrary scale-connectors w.
(d) Instead of (robustly) measuring the distance between "full probability distributions" P, Q—scaled by a "full non-probability distribution" M—with our new method one can analogously measure (robustly) the distance between the families $(P[E_z])_{z\in\mathcal{I}}$, $(Q[E_z])_{z\in\mathcal{I}}$—scaled by $(M[E_z])_{z\in\mathcal{I}}$—of probabilities of some selected concrete (e.g. increasing) events $(E_z)_{z\in\mathcal{I}} \subset \mathcal{A}$ of interest.

3 Robustness

In the previous section, we have introduced a toolbox $\mathcal{D}_{sc} := \{B_\phi(P, Q \mid M) : \phi \in \Phi, M \in \mathcal{M}_{sc}\}$ of divergences where the class of generators is (say) $\Phi = \Phi_{C_1}$ and the class $\mathcal{M}_{sc}$ of scalings consists of those measures $M = W(P, Q)$ which have λ−density $m(x) = w(p(x), q(x)) \geq 0$ with some "scale-connector" $w(\cdot, \cdot)$ between the λ−densities $p(x)$ and $q(x)$. In the following, as a part of statistical engineering, we discuss some criteria of how to find good choices of ϕ and w, having in mind robustness respectively stability respectively sensitivity (in a wide sense). For the sake of brevity, this will be done only[2] in the above-mentioned context of (finite or countable) discrete spaces $\mathcal{X} = \{x_1, x_2, \ldots\}$ of outcomes x_i where $\lambda_{count}[\{x_i\}] = 1$ for all i, $p(x) := P[\{x\}]$ and $q(x) := Q[\{x\}]$ are probability mass functions; accordingly, $w : [0, 1] \times [0, 1] \mapsto [0, \infty]$ and (5) becomes

$$\begin{aligned}
0 &\leq B_\phi\left(P, Q \mid W(P, Q)\right)\\
&= \sum_{x\in\mathcal{X}} w(p(x), q(x)) \cdot \left[\phi\left(\frac{p(x)}{w(p(x), q(x))}\right) - \phi\left(\frac{q(x)}{w(p(x), q(x))}\right)\right.\\
&\qquad \left. - \phi'\left(\frac{q(x)}{w(p(x), q(x))}\right) \cdot \left(\frac{p(x)}{w(p(x), q(x))} - \frac{q(x)}{w(p(x), q(x))}\right)\right]\\
&=: \sum_{x\in\mathcal{X}} w(p(x), q(x)) \cdot b_\phi\left(\frac{p(x)}{w(p(x), q(x))}, \frac{q(x)}{w(p(x), q(x))}\right). \qquad (13)
\end{aligned}$$

[2]The general case follows analogously and will appear elsewhere.

Since the ultimate purposes of a statistical inference based on $\mathscr{D}_{sc}$ may vary from case to case—e.g. it may serve as a basis for some desired decision/action of quite heterogeneous nature in finance, medicine, biology, signal-processing, etc.—the flexibility in choosing generators ϕ and scale-connectors w should be narrowed down in a well-structured, goal-oriented way:

- Step 1. Declare (or identify) areas A of specific interest for $(u, v) := (p(x), q(x))$: for instance, the "outlier area"

$$A_{out} := \{(u, v) \in [0, 1]^2 : u \text{ is much larger than } v, \text{ and } v \text{ is small}\} ,$$

the "inlier area"

$$A_{in} := \{(u, v) \in [0, 1]^2 : u \text{ is much smaller than } v, \text{ and } v \text{ is large}\} ,$$

the "midlier area"

$$A_{mid} := \{(u, v) \in [0, 1]^2 : u \text{ is approximately (but not exactly) equal to } v\} ,$$

and the "matching area"

$$A_{mat} := \{(v, v) : v \in [0, 1]\} .$$

The verbal names of A_{out}, A_{in} are given in accordance with applications where $P := P_N^{emp} := \frac{1}{N} \cdot \sum_{i=1}^{N} \delta_{X_i}[\cdot]$ is the above-mentioned empirical distribution of an iid sample $X_1, \ldots, X_N$ of size N, $u := p(x) = p_N^{emp}(x) = \frac{1}{N} \cdot \#\{i \in \{1, \ldots, N\} : X_i = x\}$, and $Q = P_\theta$ is a discrete candidate model distribution with $v := q(x) = p_\theta(x)$; we suggest to take this frequent situation as a running example for the robustness considerations in the whole Sect. 3. Then, A_{out} corresponds to outcomes x which appear in the sample much more often than described by the model where x is rare. Moreover, A_{in} corresponds to outcomes x which appear in the sample much less often than described by the model where x is a very frequent. In other words, A_{out}, A_{in} represent "high unusualnesses" ("surprising observations")—in terms of probabilities rather than geometry[3]—in the sampled data as compared to the candidate model. Of course, one has to define A_{out}, A_{in}, A_{mid} more quantitatively, adapted to the task to be solved. Notice that our formulation of A_{out} allows for more flexibility (e.g. expressed by its boundary) than the more restrictive "classical" definition (for some large constant $\tilde{c} > 0$ and Pearson residual $\delta := \frac{u}{v} - 1$) $A_{out}^{cl} := \{(u, v) \in [0, 1]^2 : \frac{u}{v} \geq \tilde{c}\} = \{(u, v) \in [0, 1]^2 : \delta \geq \tilde{c} - 1\} = \{(u, v) \in [0, 1]^2 : v \leq \frac{u}{\tilde{c}} \text{ if } u > 0 \text{ and } \ldots \text{ if } u = 0\}$. Analogously, our A_{in} allows for more flexibility than the classical quantification (for

[3]Which in view of the prevailing model uncertainty is a very reasonable thing; think of outliers in the lifetime data of patients infected by a lethal disease—with better modelling they may not be outliers anymore.

some small constant $\breve{c} > 0$) $A_{in}^{cl} := \{(u,v) \in [0,1]^2 : \frac{u}{v} \leq \breve{c}\} = \{(u,v) \in [0,1]^2 : \delta \leq \breve{c} - 1\} = \{(u,v) \in [0,1]^2 : v \geq \frac{u}{\breve{c}} \text{ if } u > 0 \text{ and } \ldots \text{ if } u = 0\}$; see Fig. 2l.

- Step 2. On the areas A determined in Step 1, specify desired "adjustments" on the involved generator ϕ and scale connector w which amount to dampening (downweighting) respectively amplification (highlighting) on A, to be quantified in absolute values and/or relative to a given benchmark.

In the following, we explain Step 2 in detail where for the sake of brevity we mainly concentrate on robustness (arguing on A_{out}, A_{in}); for sufficiency studies, the behaviour on A_{mid} should be focused on. To gain better insights, with the following investigations we aim for a comfortable interpretability and comparability of the underlying robustness structures, also supported through geometric 3D visualizations. To start with, we have seen in Sect. 2 that all Csiszar-Ali-Silvey divergences CASD and all classical Bregman divergences CBD are special cases of our scaled Bregman divergences SBD. Hence, we can extend all the known robustness results on CASD and CBD to a more general, unifying analysis. For the sake of brevity, this will be done only in extracts; the full details will appear elsewhere.

Case 1: Robustness without derivatives ("zeroth order"). For goodness-of-fit testing, the detection of distributional changes in data streams, two-sample tests, fast crude model search in time series (see, e.g. Kißlinger and Stummer (2015b)), and other tasks, the magnitude of $B_\phi(P, Q \mid W(P,Q))$ itself is of major importance. To evaluate its performance "microscopically", one can take an *absolute* view and inspect the magnitude of the "summand builder" (cf. (13)) $b_{\phi,w}(u,v) := w(u,v) \cdot b_\phi(\frac{u}{w(u,v)}, \frac{v}{w(u,v)})$ on the areas A of interest. For instance, depending on the data-analytic goals, one may like $b_{\phi,w}(u,v)$ to have "small (resp. large) values" on A_{out} and A_{in}. In Fig. 3 we present $b_{\phi,w_{0,1}}(u,v)$ of the following CASD (see the representation in (16) below): (a) the Kullback–Leibler divergence KL (cf. $\phi = \phi_1$), (b) the (squared) Hellinger distance HD (cf. $\phi = \phi_{1/2}$), (c) the Pearson chi-square divergence PCS (cf. $\phi = \phi_2$), as well as (d) the negative exponential disparity NED of Lindsay (1994) with $\phi(t) := \phi_{NED}(t) := \exp(1-t) + t - 2$. Concerning the non-CASD case (cf. Remark 1a above), we show in Fig. 3 $b_{\phi_2,w}(u,v)$ for the weights (e) $w = w_{no}$ (no scaling, leading to DPD_2), (f) $w = w_{0.45,\tilde{f}_6}$ (WEM scaling, cf. (10)), (g) $w = w_{0.45}^{med}$ ($\beta-$median scaling, cf. (11)), and (h) $w = w_{adj}$ with the same parameters as in Fig. 2j (cf. (12)). In (a)–(h), notice the partially large differences in the outlier area A_{out} which corresponds to the left corner. In addition to the above-mentioned *absolute* view, one can also take a *relative* view and compare the performance of the scaled Bregman divergence $B_\phi(P, Q \mid W(P,Q))$ with that of an "overall" benchmark (respectively, an alternative) $B_{\tilde{\phi}}(P, Q \mid \widetilde{W}(P,Q))$. For instance, one may like to check for $B_\phi(P, Q \mid W(P,Q)) \gtreqless B_{\tilde{\phi}}(P, Q \mid \widetilde{W}(P,Q))$, or even more "microscopically" (cf. (13)) for

$$b_{\phi,w}(u,v) = w(u,v) \cdot b_\phi\left(\frac{u}{w(u,v)}, \frac{v}{w(u,v)}\right) \gtreqless \tag{14}$$

$$\widetilde{w}(u,v) \cdot b_{\tilde{\phi}}\left(\frac{u}{\widetilde{w}(u,v)}, \frac{v}{\widetilde{w}(u,v)}\right) = b_{\tilde{\phi},\widetilde{w}}(u,v)$$

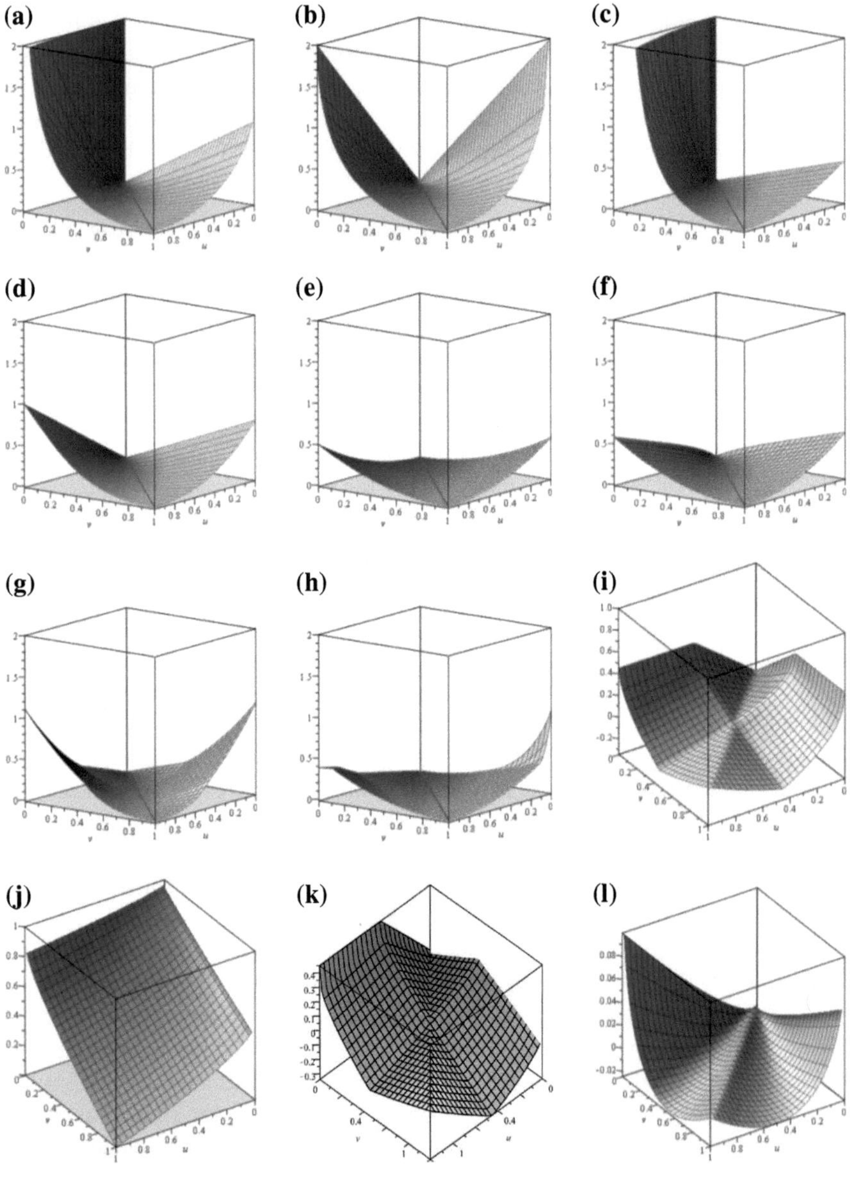

Fig. 3 Some density-pair adjustment functions DAF(0) of zeroth order (cf. (**a**)–(**h**)), and comparisons of DAF(0) (cf. (**i**)–(**l**)). **a** $b_{\phi_1, w_{0,1}}(u, v) = u \log \frac{u}{v} - (u - v)$ (KL case). **b** $b_{\phi_{1/2}, w_{0,1}}(u, v) = 2(\sqrt{u} - \sqrt{v})^2$ (HD case). **c** $b_{\phi_2, w_{0,1}}(u, v) = \frac{v}{2} \left(\frac{u}{v} - 1\right)^2$ (PCS case). **d** $b_{\phi_{NED}, w_{0,1}}(u, v) = v \cdot \phi_{NED}\left(\frac{u}{v}\right)$ (NED case). **e** $b_{\phi_2, w_{no}}(u, v) = \frac{(u-v)^2}{2}$ (DPD_2 case). **f** $b_{\phi_2, w_{0.45, \tilde{f}_6}}(u, v) = 6 \frac{(u-v)^2}{2 \log(0.45 e^{6u} + 0.55 e^{6v})}$ (cf. (10)). **g** $b_{\phi_2, w_{0.45}^{med}}(u, v) = \frac{(u-v)^2}{2 \mathrm{med}\{\min\{u,v\}, \beta, \max\{u,v\}\}}$ (cf. (11)). **h** $b_{\phi_2, w_{adj}}(u, v)$ (cf. (12)) with parameters taken as in Fig. 2j. **i** $w_{0.45}^{med}(u, v) - w_{0.45,0}(u, v) = w_{0.45}^{med}(u, v) - u^{0.45} v^{0.55}$. **j** $\mu_{w_{no}}(u, v) = 1 - \frac{0.5 v (\frac{u}{v} - 1)^2}{\frac{u}{v} \log \frac{u}{v} + 1 - \frac{u}{v}}$ (DPD_2 vs. KL case). **k** $\mu_w(u, v)$ for $w(u, v) = w_{0.45}^{med}(u, v)$. **l** $\mu_w(u, v)$ for $w(u, v) = \min(w_{0.42,1}(u, v), w_{0.25,1}(u, v))$

for all $(u, v) \in [0, 1]^2$ and especially for $(u, v) \in A_{out}, A_{in}, A_{mid}, A_{mat}$. The corresponding visual comparison can be done by overlaying $(u, v, b_{\phi,w}(u, v))$ and $(u, v, b_{\widetilde{\phi},\widetilde{w}}(u, v))$ in the same 3D plot, or by overlaying $(u, v, b_{\phi,w}(u, v) - b_{\widetilde{\phi},\widetilde{w}}(u, v))$ and $(u, v, 0)$ in another 3D plot; the latter has the advantage of an "eye-stabilizing" reference plane. Intuitively, the inequality $<$ in (14) means "point-wise" dampening (downweighting) with respect to the benchmark, whereas $>$ amounts to amplification (highlighting) with respect to the benchmark. In the light of this, one can interpret $(u, v) \mapsto b_{\phi,w}(u, v)$ as a *density-pair adjustment function DAF(0) of zeroth order*. In principle, for fixed $(\widetilde{\phi}, \widetilde{w})$ one can encounter three situations in the relative quality assessment of some candidate (ϕ, w). Firstly, the scaling is the same, i.e. $w(u, v) = \widetilde{w}(u, v)$, and thus (14) turns into

$$w(u, v) \cdot b_{\phi}\left(\frac{u}{w(u, v)}, \frac{v}{w(u, v)}\right) \gtreqless w(u, v) \cdot b_{\widetilde{\phi}}\left(\frac{u}{w(u, v)}, \frac{v}{w(u, v)}\right), \tag{15}$$

and in case of $w(u, v) > 0$ one can further simplify by dividing through $w(u, v)$. For instance, if one compares two CASD $B_{\phi}(P, Q \mid Q) \gtreqless B_{\widetilde{\phi}}(P, Q \mid Q)$ then—due to $w(u, v) = \widetilde{w}(u, v) = w_{0,1}(u, v) = v$—the inequality (15) simplifies to

$$v \cdot b_{\phi}\left(\frac{u}{v}, 1\right) = v \cdot \phi\left(\frac{u}{v}\right) \gtreqless v \cdot \widetilde{\phi}\left(\frac{u}{v}\right) = v \cdot b_{\widetilde{\phi}}\left(\frac{u}{v}, 1\right) \tag{16}$$

which on $\{(u, v) \in [0, 1]^2 : v > 0\}$ amounts to $\phi\left(\frac{u}{v}\right) \gtreqless \widetilde{\phi}\left(\frac{u}{v}\right)$. Consistently, for the outlier area one can, e.g. use the classical variant A_{out}^{cl} and for the inlier area A_{in}^{cl}. For classical unscaled Bregman divergences CBD, $B_{\phi}(P, Q \mid 1) \gtreqless B_{\widetilde{\phi}}(P, Q \mid 1)$ leads to $b_{\phi}(u, v) \gtreqless b_{\widetilde{\phi}}(u, v)$ (cf. (15)). As a second situation in comparative quality assessment, the generator functions coincide, i.e. $\phi = \widetilde{\phi}$, and hence (14) becomes

$$\begin{aligned} b_{\phi,w}(u, v) = w(u, v) \times b_{\phi}\left(\frac{u}{w(u, v)}, \frac{v}{w(u, v)}\right) &\gtreqless \\ \widetilde{w}(u, v) \times b_{\phi}\left(\frac{u}{\widetilde{w}(u, v)}, \frac{v}{\widetilde{w}(u, v)}\right) &= b_{\phi,\widetilde{w}}(u, v) . \end{aligned} \tag{17}$$

Within such a context, one can compare a CASD with its classical unscaled Bregman divergence CBD "counterpart". As an example, one can take $\phi(t) = \widetilde{\phi}(t) = \phi_2(t) = \frac{(t-1)^2}{2}$ which simplifies (17) to

$$b_{\phi_2,w}(u, v) = \frac{(u - v)^2}{2w(u, v)} \gtreqless \frac{(u - v)^2}{2\widetilde{w}(u, v)} = b_{\phi_2,\widetilde{w}}(u, v) . \tag{18}$$

The choice $w(u, v) = w_{0,1}(u, v) = v$ leads to the PCS $B_{\phi_2}(P, Q \mid Q)$ and $\widetilde{w}(u, v) = 1$ corresponds to the DPD_2 $B_{\phi_2}(P, Q \mid 1)$. Further examples of $w(u, v)$ can be drawn from Sect. 2. Clearly, (18) amounts to $w(u, v) \lesseqgtr \widetilde{w}(u, v)$ respectively $w(u, v) -$

$\widetilde{w}(u,v) \lesseqqgtr 0$ or $\frac{w(u,v)}{\widetilde{w}(u,v)} \lesseqqgtr 1$ (with special care for eventual zeros); see Fig. 3i for the difference comparison $w_{0.45}^{med}(u,v) - w_{0.45,0}(u,v)$ between the $\beta-$median BM $w_{0.45}^{med}(u,v)$ and the weighted geometric mean WGM $w_{0.45,0}(u,v)$. Notice that the case $w(u,v) = w_{adj}(u,v) \lesseqqgtr 1 = w_{no}(u,v) = \widetilde{w}(u,v)$ of the robustness adjuster against the CDB-relevant unit scaling is especially easy to compare visually. The third situation is the "crossover" case where ϕ is different from $\widetilde{\phi}$ and w is different from $\widetilde{w}$. Then, (14) needs to be treated more individually. The most fundamental example for a benchmark is clearly the Kullback–Leibler divergence KL $B_{\widetilde{\phi}}\left(P, Q \mid \widetilde{W}(P,Q)\right) = B_{\phi_1}(P, Q \mid Q)$ with generator $\widetilde{\phi}(t) = \phi_1(t) = t \log t + 1 - t \geq 0$ (with $\phi_1(1) = \phi_1'(1) = 0, \phi_1''(1) = 1$), as well as scale connector $\widetilde{w}(u,v) = w_{0,1}(u,v) = v$, yielding $b_{\widetilde{\phi},\widetilde{w}}(u,v) = v\left(\frac{u}{v}\log\left(\frac{u}{v}\right) + 1 - \frac{u}{v}\right)$. Hence, the comparison (14) specializes to

$$w(u,v) \cdot b_\phi\left(\frac{u}{w(u,v)}, \frac{v}{w(u,v)}\right) \gtreqqless v \cdot \left(\frac{u}{v}\log\left(\frac{u}{v}\right) + 1 - \frac{u}{v}\right).$$

For $\phi = \phi_2$ we get

$$\frac{(u-v)^2}{2w(u,v)} \gtreqqless v \cdot \left(\frac{u}{v}\log\left(\frac{u}{v}\right) + 1 - \frac{u}{v}\right) \iff \tag{19}$$

$$0 \gtreqqless w(u,v) - \frac{0.5 \cdot v \cdot \left(\frac{u}{v} - 1\right)^2}{\frac{u}{v} \cdot \log\left(\frac{u}{v}\right) + 1 - \frac{u}{v}} =: \mu_w(u,v)\,;$$

see, e.g. Fig. 3j for the no-scaling case $w = w_{no}$ (i.e., the DPD$_2$ vs. KL comparison) and (k) for the $\beta-$median BM case $w_{0.45}^{med}$. The check for (19) simplifies considerably if $w(u,v) = v \cdot h\left(\frac{u}{v}\right)$ for some function h (see, e.g. Fig. 3l for the scale connector $w(u,v) = \min(w_{0.42,1}(u,v), w_{0.25,1}(u,v))$ being the minimum of two weighted arithmetic means). Then the left-hand side can be rewritten as a CASD (cf. Remark 1a above) and hence context (16) applies, too.

Case 2: Robustness with first derivatives ("first order"). In many cases, one of the two distributions P, Q depends on a multidimensional parameter, say $Q \in \mathcal{Q}_\Theta := \{Q_\theta : \theta \in \Theta\}$, $\Theta \subset \mathbb{R}^d$. Then, not only the magnitude of $B_\phi(P, Q_\theta \mid W(P, Q_\theta))$ but also its derivative $\nabla_\theta\, B_\phi(P, Q_\theta \mid W(P, Q_\theta))$ with respect to θ may be of major importance. The most well-known context is minimum distance estimation, where P is a data-derived probability-measure-valued statistical functional and one wants to find the member $Q_{\widehat{\theta}}$ of $\mathcal{Q}_\Theta$ which has the shortest distance to P (if this exists), i.e. $\widehat{\theta} = \mathrm{argmin}_{\theta\in\Theta} B_\phi(P, Q_\theta \mid W(P, Q_\theta))$; for instance, $P := P_N^{emp} := \frac{1}{N} \cdot \sum_{i=1}^N \delta_{X_i}[\cdot]$ is the above-mentioned empirical distribution of an iid sample $X_1, \ldots, X_N$ of size N. If ϕ and w are smooth enough, the corresponding optimization leads to the estimating equation

$$
\begin{aligned}
0 &= -\nabla_\theta \, B_\phi \left(P, Q_\theta \mid W(P, Q_\theta)\right) \\
&= - \sum_{x \in \mathscr{X}} \nabla_\theta \, w(p(x), q_\theta(x)) \cdot b_\phi \left(\frac{p(x)}{w(p(x), q_\theta(x))}, \frac{q_\theta(x)}{w(p(x), q_\theta(x))} \right) \\
&= - \sum_{x \in \mathscr{X}} \frac{\partial}{\partial v} \left(w(p(x), v) \cdot b_\phi \left(\frac{p(x)}{w(p(x), v)}, \frac{v}{w(p(x), v)} \right) \right) \Bigg|_{v = q_\theta(x)} \cdot \nabla_\theta q_\theta(x)
\end{aligned}
$$

provided that one can interchange the sum and the derivative. In the same spirit as in Step 1, it makes sense to study the *density-pair adjustment function DAF(1) of first order* $(u, v) \mapsto a_{\phi,w}(u, v)$ defined by

$$
\begin{aligned}
a_{\phi,w}(u, v) &:= -\frac{\partial}{\partial v} b_{\phi,w}(u, v) = -\frac{\partial}{\partial v} \left\{ w(u, v) \cdot b_\phi \left(\frac{u}{w(u, v)}, \frac{v}{w(u, v)} \right) \right\} \\
&= -b_\phi \left(\frac{u}{w(u, v)}, \frac{v}{w(u, v)} \right) \cdot \frac{\partial}{\partial v} w(u, v) - w(u, v) \cdot \frac{\partial}{\partial v} b_\phi \left(\frac{u}{w(u, v)}, \frac{v}{w(u, v)} \right) \\
&= -b_\phi \left(\frac{u}{w(u, v)}, \frac{v}{w(u, v)} \right) \cdot \frac{\partial}{\partial v} w(u, v) \\
&\quad + \left\{ \phi' \left(\frac{u}{w(u, v)} \right) - \phi' \left(\frac{v}{w(u, v)} \right) \right\} \cdot \frac{u}{w(u, v)} \cdot \frac{\partial}{\partial v} w(u, v) \\
&\quad + \phi'' \left(\frac{v}{w(u, v)} \right) \cdot \left(\frac{u}{w(u, v)} - \frac{v}{w(u, v)} \right) \left\{ 1 - \frac{v}{w(u, v)} \frac{\partial}{\partial v} w(u, v) \right\} \\
&= \left\{ \frac{u}{w(u, v)} \phi' \left(\frac{u}{w(u, v)} \right) - \phi \left(\frac{u}{w(u, v)} \right) - \frac{v}{w(u, v)} \phi' \left(\frac{v}{w(u, v)} \right) + \phi \left(\frac{v}{w(u, v)} \right) \right. \\
&\quad \left. - \frac{v}{w(u, v)} \phi'' \left(\frac{v}{w(u, v)} \right) \cdot \left(\frac{u}{w(u, v)} - \frac{v}{w(u, v)} \right) \right\} \cdot \frac{\partial}{\partial v} w(u, v) \\
&\quad + \phi'' \left(\frac{v}{w(u, v)} \right) \cdot \left(\frac{u}{w(u, v)} - \frac{v}{w(u, v)} \right),
\end{aligned}
\tag{20}
$$

where we have used the definition of b_ϕ given in (13). Notice that for arbitrary CASD $B_\phi \left(P, Q \mid Q\right)$ (which corresponds to the special choice $w(u, v) = w_{0,1}(u, v) = v$), the first-order DAF(1) defined in (20) reduces by use of $\phi(1) = \phi'(1) = 0$, $C_\phi(t) := \phi(t + 1)$ $(t \in [-1, \infty[)$, and the Pearson residual $\delta = \frac{u}{v} - 1 \in [-1, \infty[$ to

$$
a_{\phi,w_{0,1}}(u, v) = \frac{u}{v} \cdot \phi' \left(\frac{u}{v} \right) - \phi \left(\frac{u}{v} \right) = (\delta + 1) \cdot C'_\phi(\delta) - C_\phi(\delta) =: \check{a}_{C_\phi}(\delta) . \tag{21}
$$

Here, $\check{a}_{C_\phi}(\cdot)$ is the *residual adjustment function* RAF of Lindsay (1994), Basu and Lindsay (1994), defined for $C_\phi-$disparities which are nothing but alternative representations of CASD divergences with generator ϕ. In other words, (21) shows that our framework of first-order density-pair adjustment functions DAF(1) $a_{\phi,w}(u, v)$ is a bivariate generalization of the concept of univariate residual adjustment functions RAFs. This extension allows in particular for comfortable *direct* comparison between the robustness structures of the (nearly disjoint) CASD and the CBD worlds, as it will be demonstrated below. Also notice that the RAF has unbounded domain, whereas the first-order DAF has bounded domain which is advantageous for plotting purposes (of bounded a). The inspection of the function $a_{\phi,w}(\cdot,\cdot)$ for the two effects dampening or amplification—especially on the areas A_{out}, A_{in}, A_{mid}, A_{mat} of interest—can be performed analogously to the inspection of $b_{\phi,w}(\cdot,\cdot)$, in absolute values and/or relative to a benchmark. For instance, for outlier and inlier dampening, $|a_{\phi,w}(\cdot,\cdot)|$ should be close to zero on A_{out}, A_{in}, and closer to zero than the benchmark. The latter may be the Kullback–Leibler divergence case

$$a_{\phi_1,w_{0,1}}(u,v) = \frac{u}{v} - 1 = \delta = \breve{a}_{C_{\phi_1}}(\delta)$$

which is positive on the area $A_{sub} := \{(u,v) \in [0,1]^2 : v < u\}$ and especially highly positive on $A_{out} \subset A_{sub}$; in contrast, $a_{\phi_1,w_{0,1}}(u,v)$ is negative on $A_{sup} := \{(u,v) \in [0,1]^2 : v > u\}$ and especially moderately negative on $A_{in} \subset A_{sup}$ (see also Fig. 21 for an illustration of A_{sub}, A_{sup} together with the classical outlier and inlier areas A^{cl}_{out}, A^{cl}_{in} which are less flexible than our A_{out}, A_{in}). In Fig. 4, we have plotted the DAF(1) $a_{\phi,w_{0,1}}(u,v)$ (cut-off at height 6) of some prominent CASD, namely of (a) the Kullback–Leibler divergence KL (cf. $\phi = \phi_1$), (b) the (squared) Hellinger distance HD (cf. $\phi = \phi_{1/2}$), (c) the Pearson chi-square divergence PCS (cf. $\phi = \phi_2$), and (d) the negative exponential disparity NED (cf. $\phi = \phi_{NED}$). Notice that on A_{sub} (and especially on the outlier area A_{out} in the left corner) the DAF(1) of HD resp. PCS resp. NED is closer to zero resp. farther from zero resp. much closer to zero than the DAF(1) of the KL; moreover, on A_{sup} (and especially on the inlier area A_{in} in the right corner) the DAF(1) of HD resp. PCS resp. NED is farther from zero resp. closer to zero resp. closer to zero than the DAF(1) of the KL (see also the corresponding difference plots (i), (j), (k) described below). This indicates that the outlier robustness of HD and NED (resp. PCS) is better (resp. worse) than that of the KL, and the inlier robustness of PCS and NED (resp. HD) is better (resp. worse) than that of the KL. Hence, these 3D plots of the bivariate DAF(1) confirm and extend the well-known results deduced from the 2D Plots of the univariate residual adjustment functions RAF. In contrast to the CASD world above, for unscaled classical Bregman divergences CBD $B_\phi(P, Q \mid 1)$ one obtains from (20) with $w(u,v) = w_{no}(u,v) = 1$ the first-order DAF(1) $a_{\phi,w_{no}}(u,v) = \phi''(v) \cdot (u - v)$. The linear dependence in u shows that all CBD have restricted flexibility in robustness modelling, compared with our SBD. To continue with other special cases, for $\phi(t) = \phi_2(t) = \frac{(t-1)^2}{2}$ we get from (20)

$$a_{\phi_2,w}(u,v) = \frac{u-v}{w(u,v)} + \frac{(u-v)^2}{2w(u,v)^2} \cdot \frac{\partial}{\partial v} w(u,v), \qquad (u,v) \in [0,1]^2. \tag{22}$$

Some examples for (22) are presented in Fig. 4, namely (e) $a_{\phi_2,w_{no}}(u,v) = u - v$ (DPD$_2$ case), (f) $a_{\phi_\alpha,w_{no}}(u,v) = (u-v) \cdot v^{\alpha-2}$ (DPD$_\alpha$ case, $\alpha = 1.67$) and (g) $a_{\phi_2,w_{\beta,\tilde{f}_r}}(u,v)$ with WEM scale connector $w_{\beta,\tilde{f}_r}(u,v)$; the inspections of the left resp. right corners indicate that the performance of (e), (g) for outliers resp. inliers are comparable and (mostly) even better than the—already very good—corresponding

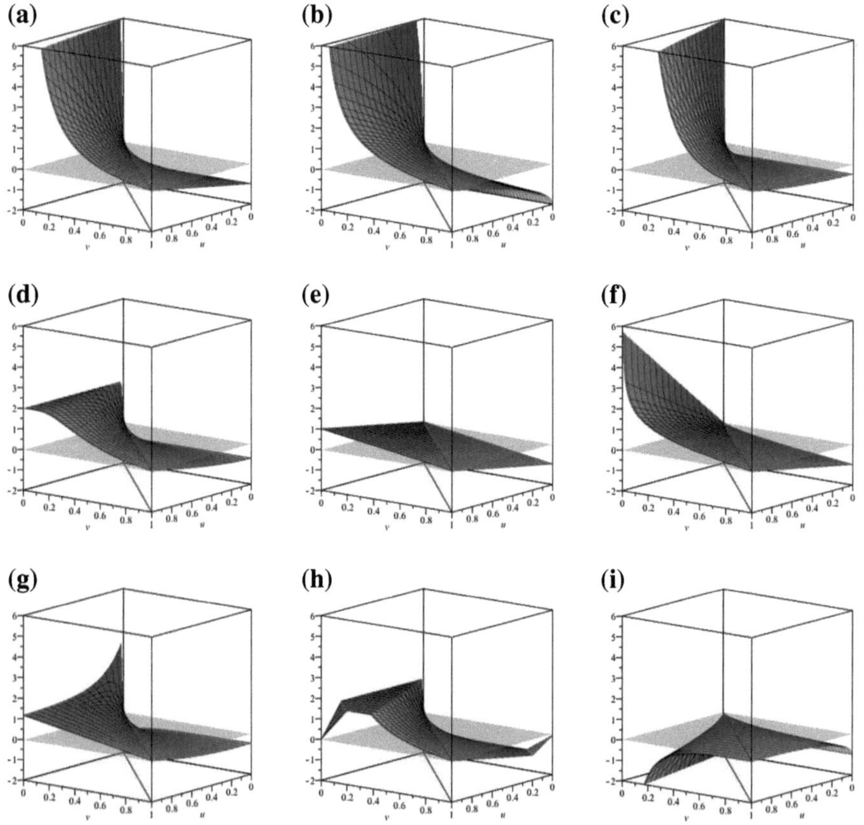

Fig. 4 Some first-order density-pair adjustment functions DAF(1) (cf. (**a**)–(**h**)), and comparisons of DAF(1) (cf. (**i**)–(**s**)). OUT refers to outlier performance (for (**a**) in absolute terms, for (**b**)–(**h**) relative to the benchmark (**a**)), IN refers to inlier performance, ASYMP refers to the asymptotics in Corollary 1 (= nice) resp. the more complicated Theorem 1 (= compl). **a** $a_{\phi_1,w_{0,1}}(u,v)=\frac{u}{v}-1$, KL (benchmark): OUT = bad, IN = ok, ASYMP= nice. **b** $a_{\phi_{1/2},w_{0,1}}(u,v)=2\cdot(\sqrt{u/v}-1)$, HD: OUT = less bad, IN = worse, ASYMP = nice. **c** $a_{\phi_2,w_{0,1}}(u,v)=((u^2/v^2)-1)/2$, PCS: OUT = worse, IN = better, ASYMP = nice. **d** $a_{\phi_{NED},w_{0,1}}(u,v)=2-(1+\frac{u}{v})\cdot\exp^{1-(u/v)}$, NED: OUT = much better, IN = better, ASYMP = nice. **e** $a_{\phi_2,w_{no}}(u,v)=u-v$, DPD$_2$: OUT = much better, IN = better, ASYMP = compl; also inefficient. **f** $a_{\phi_\alpha,w_{no}}(u,v)=(u-v)\cdot v^{\alpha-2}$, DPD$_{1.67}$: OUT = much better, IN = better, ASYMP = compl. **g** $a_{\phi_2,w_{0.45,\tilde{f}_6}}(u,v)$, $\phi=\phi_2$, $w=WEM$: OUT = much better, IN = better, ASYMP = nice. **h** $a^{des}_{adj}(u,v)$ (designed) $\phi=\phi_2$, w cf. (23): OUT = much better, IN = much better. **i** $\rho_{\phi_{1/2},w_{0,1}}(u,v)=-\left(\sqrt{\frac{u}{v}}-1\right)^2$ (HD vs. KL). **j** $\rho_{\phi_2,w_{0,1}}(u,v)=\frac{1}{2}\left(\frac{u}{v}-1\right)^2$ (PCS vs. KL). **k** $\rho_{\phi_{NED},w_{0,1}}(u,v)=2-\left(1+\frac{u}{v}\right)\exp\left(1-\frac{u}{v}\right)-\frac{u}{v}+1$ (NED vs. KL). **l** $\rho_{\phi_2,w_{0.45,\tilde{f}_6}}(u,v)$. **m** $\rho_{\phi_2,w^{des}}(u,v)=a^{des}_{adj}(u,v)-\frac{u}{v}+1$. **n** $\rho_{\phi_2,w_{no}}(u,v)=\left(\frac{u}{v}-1\right)(v-1)$ (DPD$_2$ vs. KL). **o** $a_{\phi_{1/2},w_{0,1}}(u,v)-a_{\phi_2,w_{no}}(u,v)=2\left(\sqrt{\frac{u}{v}}-1\right)-(u-v)$ (HD vs. DPD$_2$). **p** $\zeta_{w_{0,1}}(u,v)=\left(\frac{u}{v}-1\right)\cdot\left(1+\frac{1}{2}\left(\frac{u}{v}-1\right)-v\right)$ (PCS vs. DPD$_2$). **q** $a_{\phi_{NED},w_{0,1}}(u,v)-a_{\phi_2,w_{no}}(u,v)=2-\left(1+\frac{u}{v}\right)\exp\left(1-\frac{u}{v}\right)-(u-v)$ (NED vs. DPD$_2$). **r** $a_{\phi_2,w_{0.45,\tilde{f}_6}}(u,v)-a_{\phi_{NED},w_{0,1}}(u,v)$. **s** $a_{\phi_2,w_{0.45,\tilde{f}_6}}(u,v)-a_{\phi_2,w_{no}}(u,v)$

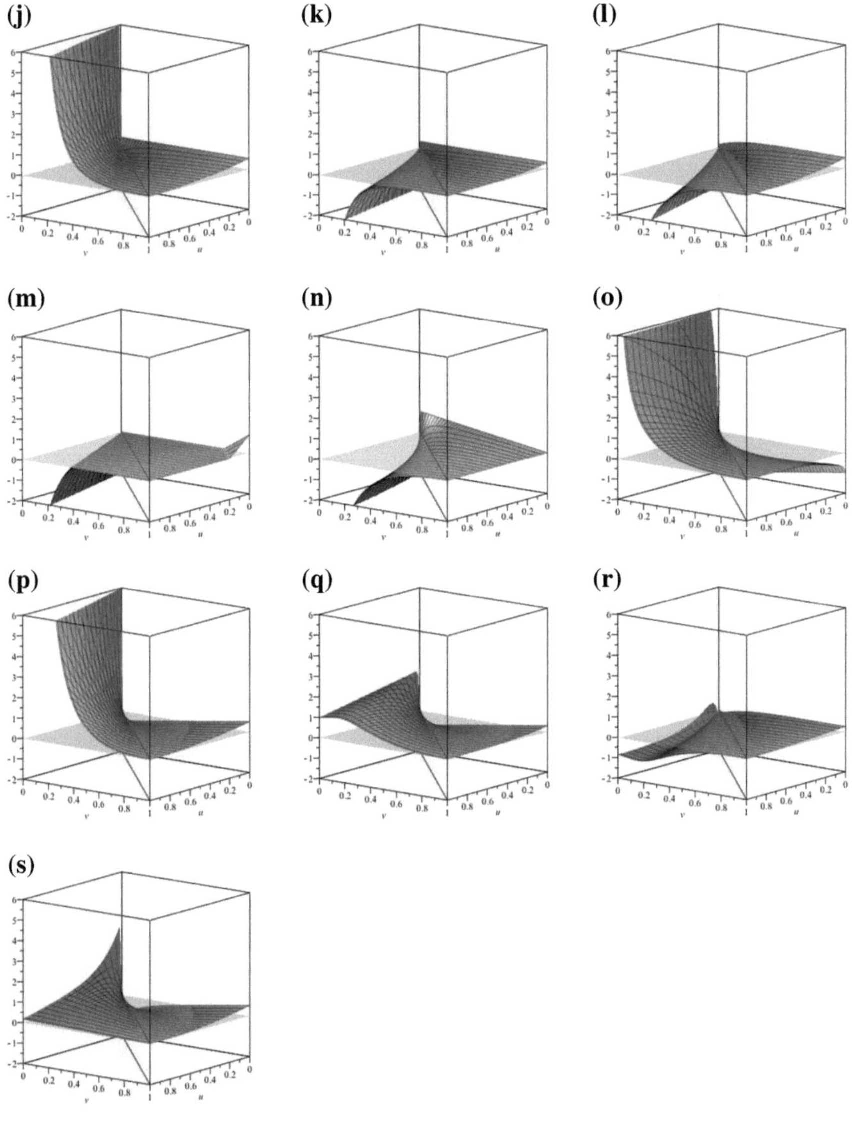

Fig. 4 (continued)

performance of the NED (d). The behaviour in the midlier area indicates that the efficiency of (g) (but not of (e)) is similar with that of (d).

One can also carry out some "reverse statistical engineering" by first fixing a first-order density-pair adjustment function $a^{des}(u, v)$ with desired properties on the areas A of interest (e.g. on A_{out}, A_{in}, A_{mid}) (and integrability with respect to $v \in [0, 1]$ should hold, too). Thereafter, one wants to deduce a corresponding scale connector

w which satisfies $a_{\phi_2,w}(u,v) = a^{des}(u,v)$ for all $(u,v) \in [0,1]^2$; this amounts to solving the differential equation produced by (22) under some designable constraint, (say), e.g. the "boundary" condition $w(\cdot, c) = h(\cdot)$ for some arbitrary strictly positive function $h(\cdot)$ on $[0,1]$ and some arbitrary constant $c \in [0,1]$. Accordingly, (with special care of possible zeros) (22) is satisfied by

$$w^{des}(u,v) = \frac{(u-v)^2}{\frac{(u-c)^2}{h(u)} + \int_c^v a^{des}(u,s)\, \mathrm{d}s}, \qquad (u,v) \in [0,1]^2 . \tag{23}$$

For instance, one can take the "first-order DAF(1) robustness adjuster"

$$\begin{aligned} a_{adj}^{des}(u,v) &:= min(u/v - 1, cut) + sgn(h_{in}) \cdot \max\left(|h_{in}| \left(1 - \frac{1}{v_{in}} - \frac{u}{u_{in}} + \frac{v}{v_{in}}\right), 0\right) \\ &+ sgn(h_{out}) \cdot \max\left(|h_{out}| \left(1 - \frac{1}{u_{out}} + \frac{u}{u_{out}} - \frac{v}{v_{out}}\right), 0\right), \end{aligned} \tag{24}$$

which is a robustified version of the first-order DAF(1) $a_{\phi_1,w_{0,1}}(u,v) = \frac{u}{v} - 1$ of the Kullback–Leibler divergence KL. The range and meaning of the constants u_{in}, v_{in}, u_{out}, v_{out}, h_{in}, h_{out} in (24) is the same as in Sect. 2.2d above (with the exception that h_{in}, h_{out} can be smaller than -1), and the cut-off constant cut is strictly positive; in Fig. 4h one can find a plot for the choice $u_{in} = v_{in} = u_{out} = v_{out} = 0.2$, $h_{in} = +0.9$, $h_{out} = -1.6$, $cut = 1.6$. As a less flexible alternative to (23), if $\varphi(u,v) := \frac{a^{des}(u,v)}{u-v} =: \varphi(v)$ does not depend on u then with the generator $\widetilde{\phi}(t) := \int_{c_1}^{v} \int_{c_2}^{z} \varphi(s)\, \mathrm{d}s$ (with arbitrary constants $c_1, c_2 \in [0,1]$) the classical Bregman divergence CBD $B_{\widetilde{\phi}}(P, Q \,|\, 1)$ has the desired DAF(1), since $a_{\widetilde{\phi},w_{no}}(u,v) = \widetilde{\phi}''(v) \cdot (u-v) = a^{des}(u,v)$.

For relative-performance analysis, we compare whether $-\nabla_\theta\ B_\phi(P, Q_\theta \,|\, W(P, Q_\theta))$ is $\gtreqless$ the benchmark $-\nabla_\theta\ B_{\widetilde{\phi}}\left(P, Q_\theta \,|\, \widetilde{W}(P, Q_\theta)\right)$. Accordingly, (20) leads to

$$a_{\phi,w}(u,v) \gtreqless a_{\widetilde{\phi},\widetilde{w}}(u,v) \iff a_{\phi,w}(u,v) - a_{\widetilde{\phi},\widetilde{w}}(u,v) \gtreqless 0 \tag{25}$$

for all $(u,v) \in [0,1] \times [0,1]$ and especially for $(u,v) \in A_{out}, A_{in}, A_{mid}, A_{mat}$. The corresponding visual comparison can be achieved by overlaying $(u, v, a_{\phi,w}(u,v))$ and $(u, v, a_{\widetilde{\phi},\widetilde{w}}(u,v))$ in the same 3D plot, or (for magnitude inspection) by overlaying $\left(u, v, \frac{a_{\phi,w}(u,v)}{a_{\widetilde{\phi},\widetilde{w}}(u,v)}\right)$ and $(u, v, 1)$ in another 3D plot. Also, the overlay of $(u, v, a_{\phi,w}(u,v) - a_{\widetilde{\phi},\widetilde{w}}(u,v))$ and $(u, v, 0)$ is a natural choice; in accordance with the above-mentioned "nearness-to-zero" robustness-quality criteria for $a_{\phi,w}$, the difference $a_{\phi,w}(u,v) - a_{\widetilde{\phi},\widetilde{w}}(u,v)$ should be negative on A_{sub} or at least on $A_{out} \subset A_{sub}$ (i.e. in the left corner of the figures) and positive on A_{sup} or at least on $A_{in} \subset A_{sup}$ (i.e. in the right corner).

For the special case $w(u,v) = \widetilde{w}(u,v) = w_{0,1}(u,v) = v$ of CASD, the corresponding 3D overlay of $(u, v, a_{\phi,w_{0,1}}(u,v))$ (cf. (21)) and $(u, v, a_{\widetilde{\phi},w_{0,1}}(u,v))$ serves as an alternative of the 2D overlay of the RAF plots $(\delta, \check{a}_{C_\phi}(\delta))$ and $(\delta, \check{a}_{C_{\widetilde{\phi}}}(\delta))$.

However, the interpretation of this kind of 3D overlay is not optimally describeable in greyscale pictures, in a few words; thus we prefer to show the a-difference-plots.

For the Kullback–Leibler divergence benchmark, with $a_{\phi_1,w_{0,1}}(u,v)) = \frac{u}{v} - 1$ we get

$$a_{\phi,w_{0,1}}(u,v) \gtreqless \frac{u}{v} - 1 = a_{\phi_1,w_{0,1}}(u,v)$$

and more generally

$$a_{\phi,w}(u,v) \gtreqless \frac{u}{v} - 1 = a_{\phi_1,w_{0,1}}(u,v) \iff \rho_{\phi,w}(u,v) := a_{\phi,w}(u,v) - \frac{u}{v} + 1 \gtreqless 0\,. \tag{26}$$

Concerning (26), in Fig. 4 we present $\rho_{\phi,w}(u,v)$ for the CASD cases of (i) (squared) Hellinger distance HD, (j) Pearson chi-square divergence PCS, and (k) negative exponential disparity NED, as well as for the non-CASD cases of (l) $\phi = \phi_2$, WEM $w = w_{\rho,\tilde{f}_r}$, (m) $\phi = \phi_2$, $w = w^{des}$ of the "first-order DAF(1) robustness adjuster" a^{des}_{adj}, and (n) $\phi = \phi_2$, $w = w_{no}$ (DPD_2 vs. KL).

Another interesting line of comparison is amongst the ϕ_2−family

$$a_{\phi_2,w}(u,v) = \frac{u-v}{w(u,v)} + \frac{(u-v)^2}{2w(u,v)^2} \cdot \frac{\partial}{\partial v} w(u,v) \gtreqless u - v = a_{\phi_2,w_{no}}(u,v)$$

$$\iff \zeta_w(u,v) := \frac{u-v}{w(u,v)} + \frac{(u-v)^2}{2w(u,v)^2} \cdot \frac{\partial}{\partial v} w(u,v) \; - (u-v) \gtreqless 0 \tag{27}$$

which for $u \neq v$ can be further simplified. Alternatively, $\frac{a_{\phi_2,w}(u,v)}{a_{\phi_2,w_{no}}(u,v)} \gtreqless 1$ is useful for the quantification of the "relative magnitude" of the method. Thus, now the benchmark is of non-CASD type, namely DPD_2. An example for the applicability of (27) can be found in Fig. 4p, where $\zeta_w(u,v)$ is given for the CASD case of Pearson's chi-square divergence PCS. In "crossover" contexts, some comparisons along the right-hand side of (25) between CASDs (other than PCS) and the DPD_2 are represented in Fig. 4o for the (squared) Hellinger distance HD, and Fig. 4q for the negative exponential disparity NED. As the NED is known to be highly robust against outliers and inliers, it makes sense to use it as a benchmark itself. In Fig. 4r we plotted $a_{\phi_2,w_{0.45,\tilde{f}_6}}(u,v) - a_{\phi_{NED},w_{0,1}}(u,v)$ which shows that our new divergence $B_{\phi_2}(P,Q \mid W_{0.45,\tilde{f}_6}(P,Q))$ has even better robustness properties than the NED which can be written in SBD form as $B_{\phi_{NED}}(P,Q \mid Q) = B_{\phi_{NED}}\big(P,Q \mid W_{0,1}(P,Q)\big)$. The comparison of the robustness of $B_{\phi_2}(P,Q \mid W_{0.45,\tilde{f}_6}(P,Q))$ with that of the DPD_2 $B_{\phi_2}(P,Q \mid 1)$ can be found in Fig. 4s by means of $a_{\phi_2,w_{0.45,\tilde{f}_6}}(u,v) - a_{\phi_2,w_{no}}(u,v)$. One can conclude that both divergences are similarly robust; however, by changing the WEM scale connector $w_{0.45,\tilde{f}_6}$ to $w_{\beta,\tilde{f}_r}$ with some other parameters $(\beta,r) \neq (0.45,6)$, one can even outperform the density power divergence DPD_2. For the sake of brevity, this will appear elsewhere.

As a final remark, let us mention that for fixed area A of interest (e.g. A_{out}, A_{in}), the behaviour of the first-order DAF(1) $a_{\phi,w}(\cdot,\cdot)$ may differ from that of the zeroth-order DAF(0) $b_{\phi,w}(\cdot,\cdot)$. For instance, the dampening (respectively, amplification) may be considerably weaker or stronger, or may even switch from dampening to amplification, or vice versa. To quantify this effect is important for avoiding undesired effects if, e.g. one uses $B_\phi\,(P, Q_\theta \mid W(P, Q_\theta))$ synchronously for minimum distance estimation and goodness-of-fit testing.

As a corresponding quantifier we suggest the *adjustment propagation function* APF

$$\eta_{\phi,w}(u,v) := \frac{a_{\phi,w}(u,v)}{b_{\phi,w}(u,v)} = -\frac{\partial}{\partial v}\log w(u,v) - \frac{\partial}{\partial v}\log b_\phi\left(\frac{u}{w(u,v)}, \frac{v}{w(u,v)}\right).$$

For example, for $\phi(t) = \phi_2(t) = \frac{(t-1)^2}{2}$ one obtains from (18) and (22) the APF $\eta_{\phi_2,w}(u,v) = \frac{2}{u-v} + \frac{\partial}{\partial v}\log w(u,v)$.

4 General Asymptotic Results for Finite Discrete Case

In this section, we assume additionally that the function $\phi(\cdot) \in \Phi_{C_1}$ is thrice continuously differentiable on $]0,\infty]$ (which implies $\phi''(1) > 0$), as well as that all three functions $w(u,v)$, $w_1(u,v) := \frac{\partial w}{\partial u}(u,v)$ and $w_{11}(u,v) := \frac{\partial^2 w}{\partial u^2}(u,v)$ are continuous in all (u,v) of some (maybe tiny) neighbourhood of the diagonal $\{(t,t) : t \in]0,1[\}$.[4] In such a setup, we deal with the following context: for $i \in \mathbb{N}$ let the observation of the ith data point be represented by the random variable X_i which takes values in some finite space $\mathscr{X} := \{x_1, \ldots, x_s\}$[5] which has $s := |\mathscr{X}| \geq 2$ outcomes (and thus, we choose $\lambda := \lambda_{count}$ as reference measure). Accordingly, let $X_1, \ldots, X_N$ represent a random sample of independent and identically distributed observations generated from an unknown true distribution $P_{\theta_{true}}$ which is supposed to be a member of a parametric family $\mathscr{P}_\Theta := \{P_\theta : \theta \in \Theta$ and P_θ has the probability mass function $p_\theta(\cdot)$ with respect to $\lambda\}$ of hypothetical, potential candidate distributions. Here, $\Theta \subset \mathbb{R}^\ell$ is a $\ell-$dimensional parameter set. Moreover, we denote by $P := P_N^{emp} := \frac{1}{N} \cdot \sum_{i=1}^N \delta_{X_i}[\cdot]$ the corresponding empirical distribution for which the probability mass function $p_N^{emp}(\cdot)$ consists of the relative frequencies $p(x) = p_N^{emp}(x) = \frac{1}{N}\cdot\#\{i \in \{1,\ldots,N\} : X_i = x\}$ (i.e. the "histogram entries"). Notice that P_N^{emp} is a probability-measure valued statistical functional (statistics). If the sample size N becomes large enough, it is intuitively plausible that the scaled Bregman divergence (cf. (13))

$$\begin{aligned} 0 \leq \frac{T_N^{\phi,w}(P_N^{emp}, P_\theta)}{2N} &:= B_\phi\left(P_N^{emp}, P_\theta \mid W(P_N^{emp}, P_\theta)\right) \\ &= \sum_{x\in\mathscr{X}} w(p_N^{emp}(x), p_\theta(x)) \cdot b_\phi\left(\frac{p_N^{emp}(x)}{w(p_N^{emp}(x), p_\theta(x))}, \frac{p_\theta(x)}{w(p_N^{emp}(x), p_\theta(x))}\right) \end{aligned}$$

[4] So that (41) holds for $\psi_{11}(u,v)$ given by (36) below.

[5] (Equipped with some $\sigma-$algebra $\mathscr{A}$).

between the data-derived empirical distribution P_N^{emp} and the candidate model P_θ converges to zero, provided that we have found the correct model in the sense that P_θ is equal to the true data generating distribution $P_{\theta_{true}}$. In the same line of argumentation, $B_\phi\left(P_N^{emp}, P_\theta \mid W(P_N^{emp}, P_\theta)\right)$ becomes close to zero, provided that P_θ is close to $P_{\theta_{true}}$. Notice that (say, for strictly positive probability mass functions $p_N^{emp}(\cdot)$ and $p_\theta(\cdot)$) the Kullback–Leibler divergence KL case

$$B_{\phi_1}\left(P_N^{emp}, P_\theta \mid W_{0,1}(P_N^{emp}, P_\theta)\right) = \sum_{x\in\mathscr{X}} p_\theta(x)\cdot\phi_1\left(\frac{p_N^{emp}(x)}{p_\theta(x)}\right) = \sum_{x\in\mathscr{X}} p_N^{emp}(x)\cdot\log\left(\frac{p_N^{emp}(x)}{p_\theta(x)}\right) \tag{28}$$

is nothing but the (multiple of the) very prominent likelihood ratio test statistics (likelihood disparity); minimizing it over θ produces the maximum likelihood estimate $\widehat{\theta}^{MLE}$. Moreover,

$$B_{\phi_2}\left(P_N^{emp}, P_\theta \mid W_{0,1}(P_N^{emp}, P_\theta)\right) = \sum_{x\in\mathscr{X}} \frac{(p_N^{emp}(x) - p_\theta(x))^2}{2p_\theta(x)}$$

represents the (multiple of the) Pearson chi-square test statistics. Concerning the above-mentioned conjectures where the sample size N tends to infinity, in case of $P_{\theta_{true}} = P_\theta$ one can even derive the *limit distribution* of the scaled-Bregman-divergence test statistics $T_N^{\phi,w}(P_N^{emp}, P_\theta)$ in quite "universal generality":

Theorem 1 *Under the null hypothesis "H_0: $P_{\theta_{true}} = P_\theta$ with $p_\theta(x) > 0$ for all $x \in \mathscr{X}$", the asymptotic distribution (as $N \to \infty$) of*

$$T_N^{\phi,w}(P_N^{emp}, P_\theta) = 2N\cdot B_\phi\left(P_N^{emp}, P_\theta \mid W(P_N^{emp}, P_\theta)\right)$$

has the following density f_{s^*}[6]*:*

$$f_{s^*}(y;\, \boldsymbol{\gamma}^{\phi,\theta}) = \frac{y^{\frac{s^*}{2}-1}}{2^{\frac{s^*}{2}}} \sum_{k=0}^{\infty} c_k \cdot \frac{\left(-\frac{y}{2}\right)^k}{\Gamma\left(\frac{s^*}{2}+k\right)}, \qquad y \in [0, \infty[, \tag{29}$$

$$\textit{with } c_0 = \prod_{j=1}^{s^*}\left(\gamma_j^{\phi,\theta}\right)^{-0.5} \textit{ and } c_k = \frac{1}{2k}\sum_{r=0}^{k-1} c_r \sum_{j=1}^{s^*}\left(\gamma_j^{\phi,\theta}\right)^{r-k} \quad (k\in\mathbb{N}) \tag{30}$$

where $s^* := rank(\boldsymbol{\Sigma A \Sigma})$ *is the number of the strictly positive eigenvalues* $(\gamma_i^{\phi,\theta})_{i=1,\ldots,s^*}$ *of the matrix* $\boldsymbol{A\Sigma} = \left(\bar{c}_i \cdot (\delta_{ij} - p_\theta(x_j))\right)_{i,j=1,\ldots,s}$ *consisting of*

[6](with respect to the one-dim. Lebesgue measure).

$$\boldsymbol{\Sigma} = (p_\theta(x_i) \cdot (\delta_{ij} - p_\theta(x_j))_{i,j=1,\dots,s} \tag{31}$$

$$\boldsymbol{A} = \left(\frac{\phi''\left(\frac{p_\theta(x_i)}{w(p_\theta(x_i),p_\theta(x_i))}\right)}{w(p_\theta(x_i),\, p_\theta(x_i))} \delta_{ij} \right)_{i,j=1,\dots,s} \tag{32}$$

$$\bar{c}_i = \phi''\left(\frac{p_\theta(x_i)}{w(p_\theta(x_i),\, p_\theta(x_i))}\right) \cdot \frac{p_\theta(x_i)}{w(p_\theta(x_i),\, p_\theta(x_i))} \,. \tag{33}$$

Here we have used Kronecker's delta δ_{ij} which is 1 iff $i = j$ and 0 else.

In particular, the asymptotic distribution (as $N \to \infty$) of $T_N^{\phi,w}(P_N^{emp}, P_\theta)$ coincides with the distribution of a weighted linear combination of standard-chi-square-distributed random variables where the weights are the $\gamma_i^{\phi,\theta}$ $(i = 1, \dots, s^*)$.

The proof of Theorem 1 is given in the appendix. Furthermore, let us mention that we can also study the asymptotics of (i) the statistics $2N \cdot B_\phi\left(P_{\widehat{\theta}_N}, P_\theta \mid W(P_{\widehat{\theta}_N}, P_\theta)\right)$ with scaled-Bregman-divergence minimum distance estimator $\widehat{\theta}_N$, as well as (ii) the two-sample statistics $2 \cdot \frac{N_1 \cdot N_2}{N_1 + N_2} \cdot B_\phi\left(P_{N_1}, P_{N_2} | W(P_{N_1}, P_{N_2})\right)$. The corresponding theorems about the latter two have a structure which is similar to that of Theorem 1, with (partially) different matrices $\mathbf{A}$ and $\boldsymbol{\Sigma}$. The details will appear elsewhere.

From the structure of Theorem 1, one can see that the asymptotic density f_{s^*} of the scaled-Bregman-divergence test statistics $2N \cdot B_\phi\left(P_N^{emp}, P_\theta \mid W(P_N^{emp}, P_\theta)\right)$ depends in general on the parameter θ. However, for a very large subclass we end up with a parameter-free chi-square limit distribution:

Corollary 1 *Let the assumptions of Theorem 1 be satisfied. If $\bar{c}_j \equiv \bar{c} > 0$ does not depend on $j \in \{1, \dots, s\}$, then $2N \cdot B_\phi\left(P_N^{emp}, P_\theta \mid W(P_N^{emp}, P_\theta)\right)/\bar{c}$ is asymptotically chi-square-distributed with $s - 1$ degrees of freedom.*

Having found out explicitly the asymptotic distribution of the test statistics, one can derive corresponding goodness-of-fit tests in a straightforward manner (chi-square-distribution quantiles, etc.). To prove Corollary 1 from Theorem 1, it is straightforward to see that $s^* = s - 1$ and $\gamma_i^{\phi,\theta} = \bar{c}$ for all $i = 1, \dots, s - 1$. Plugging this into the definitions of c_0 and c_k, one can show inductively that $c_k = \dfrac{\Gamma(k + (s^*/2))}{k! \cdot \Gamma(s^*/2) \cdot \bar{c}^{k+(s^*/2)}}$. Hence, $f_{s^*}(y;\, \boldsymbol{\gamma}^{\phi,\theta}) = \widetilde{g}(y/c)/c$ where $\widetilde{g}(\cdot)$ is the density of a chi-square distribution with s^* degrees of freedom.

The following important contexts are covered by Corollary 1:

Subcase 1: Uniform Distribution. Let the true distribution $P_{\theta_{true}}$ be the uniform distribution on $\mathscr{X} = \{1, \cdots, s\}$, i.e. under H_0 one has $p_\theta(x_i) = 1/s$ for all $i = 1, \dots, s$. Accordingly, the factor $\bar{c}_j$ becomes $\bar{c}_j = \bar{c} = \phi''\left(1/(s \cdot w\left(\frac{1}{s}, \frac{1}{s}\right))\right)/(s \cdot w\left(\frac{1}{s}, \frac{1}{s}\right))$. Hence, our Corollary 1 generalizes the Corollary 5 of Pardo and Vajda (2003) who used the uniform distribution together with the unit scaling $w(u, v) = w_{no}(u, v) = 1$.

Subcase 2: Kullback–Leibler-Divergence. For the corresponding generator $\phi(t) = \phi_1(t) = t \cdot \log(t) + 1 - t$ one gets $\phi''(t) = \frac{1}{t}$ and hence $\bar{c}_j = \bar{c} = t \cdot \phi''(t) \equiv 1$ for any arbitrary scale connector $w(\cdot,\cdot)$. Thus, Corollary 1 is much more general than the classical result that the 2N-fold of the likelihood ratio statistics (28) is asymptotically chi-square-distributed with $s-1$ degrees of freedom.

However, one can go far beyond:

Subcase 3: multiple idempotency scaling. Let $w(\cdot,\cdot)$ be an arbitrary scale connector which satisfies the condition $\exists c>0\ \forall v\in[0,\infty[\quad w(v,v)=c\cdot v$ (cf. (7)) and which is twice continuously differentiable in its first component in some neighborhood of the diagonal. Then we deduce $\bar{c}_j=\phi''\left(\frac{p_\theta(x_j)}{w(p_\theta(x_j),p_\theta(x_j))}\right)\cdot\frac{p_\theta(x_j)}{w(p_\theta(x_j),p_\theta(x_j))}=\frac{\phi''(\frac{1}{c})}{c}$. Hence, from Corollary 1 we see that for all (eventually sufficiently smoothed) scale connectors w of Sect. 2.2 and all generators ϕ, the corresponding scaled-Bregman-divergence test statistics $2N\cdot B_\phi\left(P_N^{emp},P_\theta\,|\,W(P_N^{emp},P_\theta)\right)/\bar{c}$ is asymptotically chi-square-distributed with $s-1$ degrees of freedom. From this general assertion one can immediately deduce the well-known result (see, e.g. Zografos et al. (1990), Basu and Sarkar (1994) for the one-to-one concept of disparities, Pardo (2006)) about the asymptotic chi-square-distribution (of s − 1 degrees of freedom) of all $2N/\phi''(1)$-folds of Csiszar-Ali-Silvey divergences CASD test statistics, since they are imbedded as (cf. (9) from Sect. 2.2.ai)

$$\begin{aligned}\frac{2N}{\phi''(1)}\,B_\phi\left(P_N^{emp},P_\theta\,|\,W_{0,1}(P_N^{emp},P_\theta)\right)&=\frac{2N}{\phi''(1)}\,B_\phi\left(P_N^{emp},P_\theta\,|\,P_\theta\right)\\&=\frac{2N}{\phi''(1)}\sum_{x\in\mathscr{X}}p_\theta(x)\,\phi\left(\frac{p_N^{emp}(x)}{p_\theta(x)}\right).\end{aligned}$$

To continue, notice that the case of classical-Bregman-distance test statistics

$$\begin{aligned}&2N\cdot B_\phi\left(P_N^{emp},P_\theta\,|\,1\right)=\\&\qquad 2N\sum_{x\in\mathscr{X}}\phi(P_N^{emp}(x))-\phi(p_\theta(x))-\phi'(p_\theta(x))\cdot(P_N^{emp}(x)-p_\theta(x))\end{aligned}$$

(cf. (6)) is not covered by Corollary 1; however from the more general Theorem 1 one can deduce its parameter-dependent asymptotic distribution by plugging $w(u,v)=w_{no}(u,v)=1$ into (33) which leads to $\bar{c}_j=p_\theta(x_j)\cdot\phi''\left(p_\theta(x_j)\right)$; for the special case of the DPD$_\alpha$ $2N\cdot B_{\phi_\alpha}\left(P_N^{emp},P_\theta\,|\,1\right)$ ($\alpha\in\mathbb{R}\backslash\{0,1\}$) this reduces to $\bar{c}_j=p_\theta(x_j)^{\alpha-1}$; an asymptoticity result which is similar to the latter can be found in Basu et al. (2013), with the following main differences: they use $2\alpha N\cdot B_{\phi_\alpha}(P_{\widehat{\theta}_\beta^{(N)}},P_\theta\,|\,1)$ where $\alpha>1$, P_θ need not (but is allowed to) be discrete, and $\widehat{\theta}_\beta^{(N)}$ is the parameter ϑ which minimizes $\beta\cdot B_{\phi_\beta}\left(P_N^{emp},P_\vartheta\,|\,1\right)$ for $\beta>1$.

As a modelling alternative to $2N\cdot B_\phi\left(P_N^{emp},P_\theta\,|\,1\right)$, one can work with the smoothed robustness-adjusted scale connector $w_{adj}^{smooth}(u,v)$ of Sect. 2.2d—with $h_{in}=h_{out}=0$ (see Fig. 2i) and $\underline{\varepsilon}_0=\underline{\varepsilon}_1=\overline{\varepsilon}_0=\overline{\varepsilon}_1=:\varepsilon$ with extremely small $\varepsilon>0$

(e.g. less than the rounding-errors-concerning machine epsilon of your computer); then for all practical purposes at reasonable sample sizes,

$\widetilde{B}_N := 2N\, B_\phi(P_N^{emp}, P_\theta \mid W_{adj}^{smooth}(P_N^{emp}, P_\theta))$ is "computationally indistinguishable" from $2N\, B_\phi(P_N^{emp}, P_\theta \mid 1)$, but $\widetilde{B}_N/\phi''(1)$ is asymptotically chi-square-distributed with $s-1$ degrees of freedom (being parameter-free).

Let us finally remark that all our concepts can also be performed for non-probability measures P, Q, and for similar functions. This will appear in a forthcoming paper.

Acknowledgments We are indebted to Ingo Klein for inspiring suggestions and remarks. The second author would like to thank the authorities of the Indian Statistical Institute for their great hospitality. Furthermore, we are grateful to both anonymous referees for useful suggestions.

Appendix

Proof of Theorem 1 Let us start by rewriting

$$
\begin{aligned}
&2N \cdot B_\phi\left(P_N^{emp}, P_\theta \mid W(P_N^{emp}, P_\theta)\right) \\
&= 2N \sum_{x\in\mathscr{X}} \Big[w(p_N^{emp}(x), p_\theta(x)) \cdot \Big\{ \phi\Big(\frac{p_N^{emp}(x)}{w(p_N^{emp}(x), p_\theta(x))}\Big) - \phi\Big(\frac{p_\theta(x)}{w(p_N^{emp}(x), p_\theta(x))}\Big)\Big\} \\
&-\phi'\Big(\frac{p_\theta(x)}{w(p_N^{emp}(x), p_\theta(x))}\Big) \cdot \left(p_N^{emp}(x) - p_\theta(x)\right) \Big] \\
&=: 2N \sum_{x\in\mathscr{X}} \psi(p_N^{emp}(x), p_\theta(x)),
\end{aligned}
\tag{34}
$$

where the function $\psi : [0,1]\times]0,1] \mapsto [0,\infty[$ is defined by

$$
\psi(u,v) := w(u,v)\cdot\Big\{\phi\Big(\frac{u}{w(u,v)}\Big) - \phi\Big(\frac{v}{w(u,v)}\Big)\Big\} - \phi'\Big(\frac{v}{w(u,v)}\Big)\cdot(u-v)
$$

(with the proper extension for $u=0$). As an ingredient for the below-mentioned Taylor expansion, we compute the first two partial derivatives of $\psi(\cdot,\cdot)$ with respect to its first argument:

$$
\begin{aligned}
\psi_1(u,v) &:= \frac{\partial\psi}{\partial u}(u,v) = \frac{\partial}{\partial u}\Big\{ w(u,v)\cdot\Big(\phi\Big(\frac{u}{w(u,v)}\Big) - \phi\Big(\frac{v}{w(u,v)}\Big)\Big) - \phi'\Big(\frac{v}{w(u,v)}\Big)(u-v)\Big\} \\
&= w_1(u,v)\cdot\Big(\phi\Big(\frac{u}{w(u,v)}\Big) - \phi\Big(\frac{v}{w(u,v)}\Big)\Big) \\
&\quad + w(u,v)\cdot\Big(\phi'\Big(\frac{u}{w(u,v)}\Big)\cdot\frac{w(u,v)-u\cdot w_1(u,v)}{w(u,v)^2} + \phi'\Big(\frac{v}{w(u,v)}\Big)\cdot\frac{v\cdot w_1(u,v)}{w(u,v)^2}\Big) \\
&\quad -\phi'\Big(\frac{v}{w(u,v)}\Big) + \phi''\Big(\frac{v}{w(u,v)}\Big)\cdot\frac{v\cdot w_1(u,v)}{w(u,v)^2}\cdot(u-v)
\end{aligned}
$$

$$
\begin{aligned}
&= w_1(u,v) \cdot \left(\phi\left(\frac{u}{w(u,v)}\right) - \phi\left(\frac{v}{w(u,v)}\right)\right) \\
&\quad +\phi'\left(\frac{u}{w(u,v)}\right) \cdot \frac{w(u,v) - u \cdot w_1(u,v)}{w(u,v)} - \phi'\left(\frac{v}{w(u,v)}\right) \cdot \frac{w(u,v) - v \cdot w_1(u,v)}{w(u,v)} \\
&\quad +\phi''\left(\frac{v}{w(u,v)}\right) \cdot \frac{v \cdot (u-v) \cdot w_1(u,v)}{w(u,v)^2}, \qquad \text{for all } u > 0, v > 0,
\end{aligned}
$$

$$
\psi_1(v,v) = 0 \quad \text{for all } v > 0, \tag{35}
$$

$$
\begin{aligned}
\psi_{11}(u,v) &:= \frac{\partial^2 \psi}{\partial u^2}(u,v) \\
&= w_{11}(u,v) \cdot \left(\phi\left(\frac{u}{w(u,v)}\right) - \phi\left(\frac{v}{w(u,v)}\right)\right) \\
&\quad +w_1(u,v) \cdot \left(\phi'\left(\frac{u}{w(u,v)}\right) \frac{w(u,v) - u \cdot w_1(u,v)}{w(u,v)^2} - \phi'\left(\frac{v}{w(u,v)}\right) \cdot \frac{-v \cdot w_1(u,v)}{w(u,v)^2}\right) \\
&\quad +\phi''\left(\frac{u}{w(u,v)}\right) \cdot \frac{w(u,v) - u \cdot w_1(u,v)}{w(u,v)^2} \cdot \frac{w(u,v) - u \cdot w_1(u,v)}{w(u,v)} \\
&\quad +\phi'\left(\frac{u}{w(u,v)}\right) \cdot \frac{w(u,v)\,(-u \cdot w_{11}(u,v)) - w_1(u,v) \cdot (w(u,v) - u \cdot w_1(u,v))}{w(u,v)^2} \\
&\quad -\phi''\left(\frac{v}{w(u,v)}\right) \cdot \frac{-v \cdot w_1(u,v)}{w(u,v)^2} \cdot \frac{w(u,v) - v \cdot w_1(u,v)}{w(u,v)} \\
&\quad -\phi'\left(\frac{v}{w(u,v)}\right) \cdot \frac{w(u,v) \cdot (w_1(u,v) - v \cdot w_{11}(u,v)) - (w(u,v) - v \cdot w_1(u,v)) \cdot w_1(u,v)}{w(u,v)^2} \\
&\quad +\phi'''\left(\frac{v}{w(u,v)}\right) \cdot \frac{-v \cdot w_1(u,v)}{w(u,v)^2} \cdot \frac{v \cdot (u-v) \cdot w_1(u,v)}{w(u,v)^2} \\
&\quad + \phi''\left(\frac{v}{w(u,v)}\right) \cdot \\
&\quad \frac{w(u,v)^2 \cdot (v \cdot w_1(u,v) + v \cdot (u-v) \cdot w_{11}(u,v)) - 2w(u,v) \cdot v \cdot (u-v) \cdot (w_1(u,v))^2}{w(u,v)^4} \\
&= w_{11}(u,v) \cdot \left(\phi\left(\frac{u}{w(u,v)}\right) - \phi\left(\frac{v}{w(u,v)}\right) - \frac{u}{w(u,v)} \cdot \phi'\left(\frac{u}{w(u,v)}\right)\right. \\
&\quad \left. + \frac{v}{w(u,v)} \cdot \phi'\left(\frac{v}{w(u,v)}\right) + \frac{v \cdot (u-v)}{w(u,v)^2} \cdot \phi''\left(\frac{v}{w(u,v)}\right)\right) \\
&\quad +\phi''\left(\frac{u}{w(u,v)}\right) \cdot \frac{(w(u,v) - u \cdot w_1(u,v))^2}{w(u,v)^3} \\
&\quad +\phi''\left(\frac{v}{w(u,v)}\right) \cdot \frac{w_1(u,v) \cdot (2v \cdot w(u,v) + v \cdot (v-2u) \cdot w_1(u,v))}{w(u,v)^3} \\
&\quad -\phi'''\left(\frac{v}{w(u,v)}\right) \cdot \frac{v^2 \cdot (u-v)\,(w_1(u,v))^2}{w(u,v)^4}, \qquad \text{for all } u > 0, v > 0,
\end{aligned} \tag{36}
$$

$$
\psi_{11}(v,v) = \frac{1}{w(v,v)} \cdot \phi''\left(\frac{v}{w(v,v)}\right) > 0 \qquad \text{for all } v > 0. \tag{37}
$$

By adapting the lines of Pardo and Vajda (2003) to our context, with the help of (34), (35), (37) we can perform a second-order Taylor expansion of $\psi(u, p_\theta(x))$ around $u := p_\theta(x) > 0$ ($x \in \mathcal{X}$) to achieve for each fixed sufficiently large integer N

$$
\begin{aligned}
&2N \cdot B_{\phi}\left(P_N^{emp}, P_\theta \,|\, W(P_N^{emp}, P_\theta)\right) = 2N \sum_{x\in\mathcal{X}} \psi(p_N^{emp}(x), p_\theta(x)) \\
&= 2N \cdot \sum_{x\in\mathcal{X}} \Big\{ \psi(p_\theta(x), p_\theta(x)) + \psi_1(p_\theta(x), p_\theta(x)) \cdot \left(p_N^{emp}(x) - p_\theta(x)\right) \\
&+ \frac{1}{2}\, \psi_{11}(p_\theta^*(x), p_\theta(x)) \cdot \left(p_N^{emp}(x) - p_\theta(x)\right)^2 \Big\} \\
&= N \cdot \sum_{x\in\mathcal{X}} \psi_{11}(p_\theta^*(x), p_\theta(x)) \cdot \left(p_N^{emp}(x) - p_\theta(x)\right)^2
\end{aligned}
$$

for some $p_\theta^*(x)$ with $|p_\theta^*(x) - p_\theta(x)| \leq |p_N^{emp}(x) - p_\theta(x)|$ $(x \in \mathcal{X})$. Therefore, by using (37) we obtain

$$
\begin{aligned}
&\left| 2N \cdot B_{\phi}\left(P_N^{emp}, P_\theta \,|\, W(P_N^{emp}, P_\theta)\right) - N \cdot \sum_{x\in\mathcal{X}} \frac{\phi''\left(\frac{p_\theta(x)}{w(p_\theta(x), p_\theta(x))}\right)}{w(p_\theta(x), p_\theta(x))} \cdot \left(p_N^{emp}(x) - p_\theta(x)\right)^2 \right| \\
&\leq N \cdot \sum_{x\in\mathcal{X}} \left| \psi_{11}(p_\theta^*(x), p_\theta(x)) - \psi_{11}(p_\theta(x), p_\theta(x)) \right| \cdot \left(p_N^{emp}(x) - p_\theta(x)\right)^2 \\
&= N \cdot \sum_{x\in\mathcal{X}} \left| \frac{\psi_{11}(p_\theta^*(x), p_\theta(x)) - \psi_{11}(p_\theta(x), p_\theta(x))}{\psi_{11}(p_\theta(x), p_\theta(x))} \right| \cdot \psi_{11}(p_\theta(x), p_\theta(x)) \cdot \left(p_N^{emp}(x) - p_\theta(x)\right)^2 \\
&\leq \left\{ \sup_{(u,v): |u-v| \leq \sup_{x\in\mathcal{X}} |p_N^{emp}(x) - p_\theta(x)|} \left| \frac{\psi_{11}(u,v)}{\psi_{11}(v,v)} - 1 \right| \right\} \cdot N \cdot \\
&\qquad \sum_{x\in\mathcal{X}} \frac{\phi''\left(\frac{p_\theta(x)}{w(p_\theta(x), p_\theta(x))}\right)}{w(p_\theta(x), p_\theta(x))} \cdot \left(p_N^{emp}(x) - p_\theta(x)\right)^2 .
\end{aligned} \tag{38}
$$

Denoting the random (overall) sup −term within the curly brackets of (38) by Y_N one gets from (38) the sandwich (squeeze) bounds

$$
\begin{aligned}
&\sup \left\{ 0, (1 - Y_N) \cdot N \cdot \sum_{x\in\mathcal{X}} \frac{\phi''\left(\frac{p_\theta(x)}{w(p_\theta(x), p_\theta(x))}\right)}{w(p_\theta(x), p_\theta(x))} \cdot \left(p_N^{emp}(x) - p_\theta(x)\right)^2 \right\} \\
&\leq 2N \cdot B_{\phi}\left(P_N^{emp}, P_\theta \,|\, W(P_N^{emp}, P_\theta)\right) \\
&\leq (1 + Y_N) \cdot N \cdot \sum_{x\in\mathcal{X}} \frac{\phi''\left(\frac{p_\theta(x)}{w(p_\theta(x), p_\theta(x))}\right)}{w(p_\theta(x), p_\theta(x))} \cdot \left(p_N^{emp}(x) - p_\theta(x)\right)^2 .
\end{aligned} \tag{39}
$$

But

$$
Y_N \xrightarrow{a.s.} 0 \quad \text{as } N \to \infty, \tag{40}
$$

since $\sup_{x\in\mathcal{X}} |p_N^{emp}(x) - p_\theta(x)| \xrightarrow{a.s.} 0$ as $N \to \infty$ (cf. e.g. Vapnik and Chervonenkis (1968)) and

$$\lim_{c \to 0_+} \sup_{(u,v):|u-v| \leq c} \left| \frac{\psi_{11}(u,v)}{\psi_{11}(v,v)} - 1 \right| = 0 \tag{41}$$

due to our assumptions on the functions $\phi(\cdot)$ and $w(\cdot,\cdot)$. Furthermore, it is well known (cf. e.g. Serfling 1980) that the $N-$sequence of random vectors

$$R_N := \sqrt{N} \cdot \left(p_N^{emp}(x_1) - p_\theta(x_1), \ldots, p_N^{emp}(x_s) - p_\theta(x_s)\right)^{tr}$$

converges in distribution to a s-variate normal R with mean vector 0 and covariance matrix Σ given by (31). Thus, in terms of the matrix A given by (32) one gets the convergence in distribution

$$N \cdot \sum_{x \in \mathscr{X}} \frac{\phi''\left(\frac{p_\theta(x)}{w(p_\theta(x), p_\theta(x))}\right)}{w(p_\theta(x), p_\theta(x))} \cdot \left(p_N^{emp}(x) - p_\theta(x)\right)^2 = R_N^{tr} \cdot A \cdot R_N \xrightarrow[N\to\infty]{\mathscr{D}} R^{tr} \cdot A \cdot R =: \widetilde{R}. \tag{42}$$

In addition, one can apply a theorem of Dik and de Gunst (1985) on quadratic forms of normal variables to deduce that the distribution of $\widetilde{R}$ coincides with that of $\breve{R} := \sum_{j=1}^{s^*} \gamma_j^{\phi,\theta} \cdot Z_j^2$, where $s^* := rank(\mathbf{\Sigma A \Sigma})$ is the number of the (ordered) eigenvalues $\gamma_1^{\phi,\theta} \geq \gamma_2^{\phi,\theta} \geq \ldots \geq \gamma_{s^*}^{\phi,\theta} > 0$ of the matrix $\mathbf{A\Sigma}$ and $Z_1, \ldots, Z_{s^*}$ are i.i.d. (univariate) standard normal random variables. The corresponding power series expansion of the density of $\breve{R}$ follows from investigations of Kotz et al. (1967), which we slightly simplify in the following. For the case $s^* = 1$ we have $\breve{R} = \gamma_1^{\phi,\theta} \cdot Z_1^2$ with density $g(y) := (2\pi \cdot \gamma_1^{\phi,\theta} \cdot y)^{-1/2} \cdot \exp\left(-\frac{y}{2\gamma_1^{\phi,\theta}}\right)$; on the other hand, by straightforward induction one can explicitly solve (30) to $c_k = \frac{(2k)!}{(k!)^2 \cdot 4^k \cdot (\gamma_1^{\phi,\theta})^{k+1/2}}$ $(k \in \mathbb{N}_0)$, and plugging this into (29) leads to the desired representation $f_{s^*}(y;\, \gamma_1^{\phi,\theta}) = g(y)$. In the remaining case $s^* \in \mathbb{N}\backslash\{1\}$ (and thus, $\frac{s^*}{2} - 1 \geq 0$), for complex-valued t with $Re(t) > -(2\gamma_1^{\phi,\theta})^{-1}$, $\widetilde{c_0} := \prod_{j=1}^{s^*} \left(\gamma_j^{\phi,\theta}\right)^{-0.5}$ and $d_k := \frac{1}{2} \sum_{j=1}^{s^*} \left(\gamma_j^{\phi,\theta}\right)^{-k}$ $(k \in \mathbb{N})$ let us compute the "modified logarithmic Laplace transform"

$$\log\left\{(2t)^{s^*/2} \cdot E\left[\exp\left(-t \cdot \breve{R}\right)\right]\right\} = \log\left\{(2t)^{s^*/2} \cdot \prod_{j=1}^{s^*} E\left[\exp\left(-t \cdot \gamma_j^{\phi,\theta} \cdot Z_j^2\right)\right]\right\}$$

$$= \log\left\{(2t)^{s^*/2} \cdot \prod_{j=1}^{s^*} \left(1 + 2t \cdot \gamma_j^{\phi,\theta}\right)^{-1/2}\right\} = \log(\widetilde{c_0}) + \sum_{j=1}^{s^*} \left(-\frac{1}{2}\right) \cdot \log\left(1 + \frac{1}{2t \cdot \gamma_j^{\phi,\theta}}\right)$$

$$= \log(\widetilde{c_0}) + \sum_{j=1}^{s^*} \sum_{k=1}^{\infty} \frac{1}{2k} \cdot \left(-\frac{1}{2t \cdot \gamma_j^{\phi,\theta}}\right)^k = \log(\widetilde{c_0}) + \sum_{k=1}^{\infty} \frac{d_k}{k} \cdot \left(-\frac{1}{2t}\right)^k =: M_1\left(-(2t)^{-1}\right). \tag{43}$$

On the other hand, for the density $g(\cdot)$ of $\breve{R}$ one can make the ansatz

$$g(y) := \sum_{k=0}^{\infty} c_k \cdot h_k(y), \qquad y \in [0, \infty[, \tag{44}$$

for some sequence $(c_k)_{k\in\mathbb{N}_0}$ of strictly positive real numbers and the power functions $h_k(y) := \dfrac{\left(\frac{y}{2}\right)^{s^*/2-1}\cdot\left(-\frac{y}{2}\right)^k}{2\Gamma\left(\frac{s^*}{2}+k\right)}$ (recall that $\frac{s^*}{2}-1\geq 0$). The corresponding modified logarithmic Laplace transform of g computes—formally—as

$$\log\left\{(2t)^{s^*/2}\cdot\int_{[0,\infty[}\exp(-t\cdot y)\cdot g(y)\,\mathrm{d}y\right\} = \log\left\{(2t)^{s^*/2}\cdot\int_{[0,\infty[}\exp(-t\cdot y)\cdot\sum_{k=0}^{\infty}c_k\cdot h_k(y)\,\mathrm{d}y\right\}$$
$$= \log\left\{\sum_{k=0}^{\infty}c_k\cdot(2t)^{s^*/2}\cdot\int_{[0,\infty[}\exp(-t\cdot y)\cdot h_k(y)\,\mathrm{d}y\right\}$$
$$= \log\left\{\sum_{k=0}^{\infty}c_k\cdot\left(-\frac{1}{2t}\right)^k\right\} =: M_2\left(-(2t)^{-1}\right). \tag{45}$$

By the uniqueness of the Laplace transform, the two functions $M_1(\cdot)$ and $M_2(\cdot)$ have to coincide, and thus, by means of the reparametrization $\chi := -(2t)^{-1}$ the coefficients c_k $(k\geq 1)$ can be identified by equating their derivatives which implies

$$\sum_{k=0}^{\infty}c_{k+1}\cdot(k+1)\cdot\chi^k = \left(\sum_{k=0}^{\infty}c_k\cdot\chi^k\right)\cdot\left(\sum_{k=0}^{\infty}d_{k+1}\cdot\chi^k\right) = \sum_{k=0}^{\infty}\left(\sum_{r=0}^{k}c_r\cdot d_{k-r+1}\right)\cdot\chi^k. \tag{46}$$

Hence, $c_{k+1} = \sum_{r=0}^{k}c_r\cdot d_{k+1-r}$ for all $k\in\mathbb{N}_0$ and also $c_0 = \widetilde{c_0}$ (from $M_1(0) = M_2(0)$), which together with (44) leads to the desired series expansion (29) of the density of $\breve{R}$. Finally, the interchangement in the second equality of (45) can be justified by Lebesgue's dominated convergence theorem, since

$$c_0\cdot\sum_{k=0}^{\infty}\frac{\chi^k}{k!}\cdot E\left[\left(\sum_{j=1}^{s^*}\frac{Z_j^2}{2\gamma_j^{\phi,\theta}}\right)^k\right] = c_0\cdot\prod_{j=1}^{s^*}E\left[\exp\left(\frac{\chi}{2\gamma_j^{\phi,\theta}}\cdot Z_j^2\right)\right] = c_0\cdot\prod_{j=1}^{s^*}\left(1-\frac{2\chi}{2\gamma_j^{\phi,\theta}}\right)^{-1/2}$$
$$= \exp\left(M_1(\chi)\right) = \exp\left(M_2(\chi)\right) = \sum_{k=0}^{\infty}c_k\cdot\chi^k$$

and thus for all $y>0$

$$\sum_{k=0}^{\infty}|c_k|\cdot|h_k(y)| = \sum_{k=0}^{\infty}\frac{c_0}{k!}\cdot E\left[\left(\sum_{j=1}^{s^*}\frac{Z_j^2}{2\gamma_j^{\phi,\theta}}\right)^k\right]\cdot|h_k(y)| \leq \sum_{k=0}^{\infty}\frac{c_0}{k!\cdot(2\gamma_{s^*}^{\phi,\theta})^k}\cdot E\left[\left(\sum_{j=1}^{s^*}Z_j^2\right)^k\right]\cdot|h_k(y)|$$
$$= \sum_{k=0}^{\infty}\frac{c_0}{k!\cdot(2\gamma_{s^*}^{\phi,\theta})^k}\cdot\frac{2^k\cdot\Gamma\left(k+\frac{s^*}{2}\right)}{\Gamma\left(\frac{s^*}{2}\right)}\cdot\frac{\left(\frac{y}{2}\right)^{s^*/2+k-1}}{2\Gamma\left(\frac{s^*}{2}+k\right)} = \frac{c_0}{2^{s^*/2}\cdot\Gamma\left(\frac{s^*}{2}\right)}\cdot y^{s^*/2-1}\cdot\exp\left(\frac{y}{2\gamma_{s^*}^{\phi,\theta}}\right).$$

□

References

Ali SM, Silvey SD (1966) A general class of coefficients of divergence of one distribution from another. J Roy Stat Soc B 28:131–142

Basu A, Lindsay BG (1994) Minimum disparity estimation for continuous models: efficiency, distributions and robustness. Ann Inst Statis Math 46:683–705

Basu A, Sarkar S (1994) On disparity based goodness-of-fit tests for multinomial models. Statis Probab Lett 19:307–312

Basu A, Harris IR, Hjort N, Jones M (1998) Robust and efficient estimation by minimising a density power divergence. Biometrika 85(3):549–559

Basu A, Shioya H, Park C (2011) Statistical inference: the minimum distance approach. CRC, Boca Raton

Basu A, Mandal A, Martin N, Pardo L (2013) Testing statistical hypotheses based on the density power divergence. Ann Inst Statis Math 65(2):319–348

Basu A, Mandal A, Martin N, Pardo L (2015a) Density power divergence tests for composite null hypotheses. arXiv:14030330v2

Basu A, Mandal A, Martin N, Pardo L (2015b) Robust tests for the equality of two normal means based on the density power divergence. Metrika 78:611–634

Beran RJ (1977) Minimum hellinger distance estimates for parametric models. Ann Stat 5:445–463

Bregman LM (1967) The relaxation method of finding the common point of convex sets and its application to the solution of problems in convex programming. USSR Comput Math Math Phys 7(3):200–217

Csiszar I (1963) Eine informationstheoretische Ungleichung und ihre Anwendung auf den Beweis der Ergodizität von Markoffschen Ketten. Publ Math Inst Hungar Acad Sci A 8:85–108

Csiszar I (1991) Why least squares and maximum entropy? An axiomatic approach to inference for linear inverse problems. Ann Stat 19(4):2032–2066

Csiszar I (1994) Maximum entropy and related methods. In: Transactions 12th Prague Conference Information Theory, Statistical Decision Functions and Random Processes, Czech Acad Sci Prague, pp 58–62

Csiszar I (1995) Generalized projections for non-negative functions. Acta Mathematica Hungarica 68:161–186

Csiszar I, Shields PC (2004) Information theory and statistics: a tutorial. now. Hanover, Mass

Dik JJ, de Gunst MCM (1985) The distribution of general quadratic forms in normal variables. Statistica Neerlandica 39:14–26

Ghosh A, Basu A (2013) Robust estimation for independent non-homogeneous observations using density power divergence with applications to linear regression. Electron J Stat 7:2420–2456

Ghosh A, Basu A (2014) Robust and efficient parameter etsimation based on censored data with stochastic covariates. arXiv:14105170v2

Golan A (2003) Information and entropy econometrics editors view. J Econometrics 107:1–15

Grabisch M, Marichal JL, Mesiar R, Pap E (2009) Aggregation functions. Cambridge University Press

Kißlinger AL, Stummer W (2013) Some decision procedures based on scaled Bregman distance surfaces. In: Nielsen F, Barbaresco F (eds) GSI 2013, Lecture Notes in Computer Science LNCS, 8085. Springer, Berlin, pp 479–486

Kißlinger AL, Stummer W (2015a) A new information-geometric method of change detection. Preprint

Kißlinger AL, Stummer W (2015b) New model search for nonlinear recursive models, regressions and autoregressions. In: Nielsen F, Barbaresco F, SCSL (eds) GSI 2015, Lecture Notes in Computer Science LNCS 9389. Springer, Switzerland, pp 693–701

Kotz S, Johnson N, Boyd D (1967) Series representations of distributions of quadratic forms in normal variables. i. central case. Ann Math Stat 38(3):823–837

Liese F, Miescke KJ (2008) Statistical Decision Theory: Estimation, Testing, and Selection. Springer, New York

Liese F, Vajda I (1987) Convex statistical distances. Teubner, Leipzig
Liese F, Vajda I (2006) On divergences and informations in statistics and information theory. IEEE Trans Inf Theory 52(10):4394–4412
Lindsay BG (1994) Efficiency versus robustness: the case for minimum Hellinger distance and related methods. Ann Statis 22(2):1081–1114
Maasoumi E (1993) A compendium to information theory in economics and econometrics. Econometrics Rev 12(2):137–181
Marhuenda Y, Morales D, Pardo JA, Pardo MC (2005) Choosing the best Rukhin goodness-of-fit statistics. Comp Statis Data Anal 49:643–662
Pardo L (2006) Statistical inference based on divergence measures. Chapman & Hall/CRC, Taylor & Francis Group
Pardo MC, Vajda I (1997) About distances of discrete distributions satisfying the data processing theorem of information theory. IEEE Trans Inf Theory 43(4):1288–1293
Pardo MC, Vajda I (2003) On asymptotic properties of information-theoretic divergences. IEEE Trans Inf Theory 49(7):1860–1868
Read TRC, Cressie NAC (1988) Goodness-of-fit statistics for discrete multivariate data. Springer, New York
Rukhin AL (1994) Optimal estimator for the mixture parameter by the method of moments and information affinity. In: Transactiona 12th Prague Conference Information Theory, Statistical Decision Functions and Random Processes. Czech Acad Sci, Prague, pp 214–216
Serfling RJ (1980) Approximation theorems of mathematical statistics. Wiley Series in Probability and Mathematical Statistics
Stummer W (2004) Exponentials, diffusions, finance, entropy and information. Shaker, Aachen
Stummer W (2007) Some Bregman distances between financial diffusion processes. Proc Appl Math Mech (PAMM) 7:1050,503–1050,504
Stummer W, Lao W (2012) Limits of Bayesian decision related quantities of binomial asset price models. Kybernetika 48(4):750–767
Stummer W, Vajda I (2007) Optimal statistical decisions about some alternative financial models. J Econometrics 137:441–471
Stummer W, Vajda I (2012) On Bregman distances and divergences of probability measures. IEEE Trans Inf Theory 58(3):1277–1288
Vajda I (1989) Theory of statistical inference and information. Kluwer, Dordrecht
Vajda I, van der Meulen EC (2010) Goodness-of-fit criteria based on observations quantized by hypothetical and empirical percentiles. In: Karian Z, Dudewicz E (eds) Handbook of Fitting statistical distributions with R. CRC, Heidelberg, pp 917–994
Vapnik VN, Chervonenkis AY (1968) On the uniform convergence of frequencies of occurence of events to their probabilities. Sov Math Doklady 9(4):915–918, corrected reprint in: Schölkopf B et al (eds) (2013) Empirical Inference. Springer, Berlin, pp 7–12
Voinov V, Nikulin M, Balakrishnan N (2013) Chi-squared goodness of fit tests with applications. Academic Press
Zografos K, Ferentinos K, Papaioannou T (1990) Phi-divergence statistics: sampling properties and multinomial goodness of fit and divergence tests. Commun Statist A - Theory Meth 19(5):1785–1802

SB-Robustness of Estimators

Arnab Kumar Laha and A.C. Pravida Raja

1 Introduction

Outliers are observations which are strikingly different from the rest in a dataset. Real-life datasets often contain a few outliers. Many commonly used statistical procedures are substantially impacted by the presence of outliers in a dataset which is undesirable. However, robust statistical procedures are not significantly impacted by the presence of outliers and hence are more suitable for use in situations where one suspects that outliers may be present in the dataset. In showing how an estimator responds to the introduction of a new observation, Hampel (1968, 1974) introduced the influence curve (IC) a.k.a influence function (IF) which allows us to understand the relative influence of individual observations on the value of an estimate or test statistic.

Huber (1964) introduced the gross error model $F(x-\theta) = (1-\varepsilon)G(x-\theta) + \varepsilon H(x-\theta)$ assuming that a known fraction ε, $(0 \leq \varepsilon < 1)$ of the data may consist of "gross errors" with an arbitrary unknown distribution $H(x-\theta)$ while the rest of the data come from a parametric model $G(x-\theta)$ for known G. The influence function of the functional T at the underlying distribution F is defined as

$$IF(x; T, F) = \lim_{\varepsilon \to 0} \frac{T((1-\varepsilon)F + \varepsilon\delta_x) - T(F)}{\varepsilon}$$

where δ_x denote the degenerate distribution assigning probability one to the point x.

The gross error sensitivity (g.e.s.) of the estimator T at F is defined as (Hampel 1974)

A.K. Laha (✉)
Production and Quantitative Methods, Indian Institute of Management Ahmedabad, Ahmedabad 380015, Gujarat, India
e-mail: arnab@iima.ac.in

A.C. Pravida Raja
Indian Institute of Management Ahmedabad, Ahmedabad 380015, Gujarat, India
e-mail: pravida@iima.ac.in

C. Agostinelli et al. (eds.), *Recent Advances in Robust Statistics: Theory and Applications*, DOI 10.1007/978-81-322-3643-6_6

$$\gamma(T, F) = sup_x |IF(x; T, F)|$$

If $\gamma(T, F)$ is finite then the estimator is said to be bias-robust (or B-robust) at F (Rousseeuw 1981).

When working with bounded parameter spaces or bounded support the notion of robustness based on finiteness of g.e.s. needs modification. This is because g.e.s is bounded in such situations. Ko and Guttorp (1988) introduced the notion of Standardized Influence Function (SIF) of a functional T w.r.t. a functional F as

$$SIF(x; T, F, S) = \frac{|IF(x; T, F)|}{S(F)}, S(F) \neq 0$$

where F is the underlying distribution. The standardized gross error sensitivity (s.g.e.s) of T with respect to the functional S at the family of distributions $\Im$ is defined as

$$\gamma^*(T, \Im, S) = sup_{\Im} sup_x SIF(x; T, F, S)$$

If $\gamma^*(T, \Im, S)$ is finite then the estimator is said to be standardized bias robust (or SB-robust) at the family of distributions $\Im$. It may be noted that the notion of SB-robustness depends on the choice of the functional S. Usually, S is taken to be a dispersion measure and hence the notion of SB-robustness depends on the choice of the dispersion measure used. Ko and Guttorp (1988) gives a set of desirable conditions that a measure of dispersion S on a (q-1)-dimensional sphere Ω_q of $\Re^q$ should satisfy. Let X and Y be two random unit vectors with unimodal distributions F and G with modal vectors $T(X)$ and $T(Y)$ respectively. A real-valued functional S is called a dispersion on Ω_q,

1. $S(F) \leq$ S(G) whenever $d(Y, T(Y))$ is stochastically larger than $d(X, T(X))$ where d is a metric on Ω_q
2. $S(F) =$ S(G) if $Y = \Gamma(X)$ for an orthogonal matrix Γ
3. $S(\delta_c) = 0$ if c is a fixed point on Ω_q.

Circular data analysis differs from the standard univariate or multivariate data analysis in that it requires the inference not to depend on the choice of origin and sense of rotation. For this reason, the arithmetic mean as well as standard deviation are not useful as measures of central tendency and dispersion when working with circular data. Treating each angular observation Θ as a unit vector joining the origin with the point (cos Θ, sin Θ), we define the mean direction of a set of angular observations as the direction of their resultant vector.

Let $C = \sum_1^n \cos \Theta_i$ and $S = \sum_1^n \sin \Theta_i$ where $\Theta_i^{\prime}s$ are independently and identically distributed circular random variables. The circular mean direction of a set of n angular observations Θ_1,Θ_2,...,Θ_n is given by

$$\bar{\Theta}_0 = arctan^*(\frac{S}{C})$$

where $arctan^*$ is the quadrant-specific inverse of the tangent function which is defined as

$$arctan^*(\frac{S}{C}) = \begin{cases} arctan(\frac{S}{C}) & \text{if } C > 0, S \geq 0 \\ \frac{\pi}{2} & \text{if } C = 0, S > 0 \\ arctan(\frac{S}{C}) + \pi & \text{if } C < 0 \\ arctan(\frac{S}{C}) + 2\pi & \text{if } C \geq 0, S < 0 \end{cases}$$

(see Jammalamadaka and SenGupta (2001), p. 13). When both $C = 0$ and $S = 0$ the circular mean direction is not defined. Thus, for the uniform distribution on the unit circle, the circular mean direction is not defined. It should be noted that the circular mean direction does not depend on the choice of origin and the sense of rotation. That is, it is rotationally invariant. The length of the resultant vector $R = \sqrt{C^2 + S^2}$ is a useful measure of concentration for unimodal data. If R is close to 0 then dispersion is large whereas a value of R close to n imply that the observations have small dispersion or more concentration toward the mean direction.

The outlier problem in the circular data setup differs considerably from that in the univariate linear data case. Since there is not much room for an observation to out lie it may be expected that fewer outlier problems will arise in circular data analysis. Moreover, the presence of outliers can be detected only if the remaining observations have high concentration. Jammalamadaka and SenGupta (2001) suggests use of an appropriate circular distance to judge how far an observation is from the circular mean for identifying outliers. The robustness properties of statistical procedures for circular data have not been much studied (see however Mardia and Jupp (2000), pp. 267–269).

Wehrly and Shine (1981) derived the influence function of the circular mean $T(F) = arctan^*[\frac{E_F(sin\Theta)}{E_F(cos\Theta)}]$ as $IF(\theta; T, F) = \frac{sin(\theta - \mu_F)}{\rho_F}$ where μ_F and ρ_F are the mean direction and concentration parameter of the underlying distribution. For any value of θ, the influence curve is bounded by $\pm\rho_F^{-1}$. Thus, the circular mean is B-robust. Laha and Mahesh (2011) proved that the circular mean $T(F)$ is B-robust but not SB-robust at the family of distributions $\Im = \{vM(\mu, \kappa), \kappa > 0\}$ when the measure of dispersion is $S(F) = E_F(d(\Theta, \mu))$ where $vM(\mu, \kappa)$ denotes the von-Mises distribution with mean direction μ and concentration parameter κ (see Sect. 2 for the definition), $F \in \Im$ and $d(\Theta, \mu) = min(\Theta - \mu, 2\pi - (\Theta - \mu))$.

An alternative approach to the robustness problem for estimators in the circular data set-up is given in Agostinelli (2007). He introduced robust estimators based on minimum disparity measures (MDE) and weighted likelihood estimating equations (WLE). Since outliers are observations which are highly unlikely to occur under the assumed model, these estimators are based on the difference of the estimated density (using a kernel density estimator) from the assumed model. In the absence of outliers, the estimator based on weighted likelihood is asymptotically equivalent to the maximum likelihood estimator but has positive breakdown point in the presence of outliers. However, both these estimators are not SB-robust.

He and Simpson (1992) introduced distance-based breakdown function for general parametric family of distributions and proposed an alternative definition of SB-robustness based on the same. They show that under certain conditions the notion of SB-robustness proposed by them coincides with that of Ko and Guttorp (1988) which we have adapted for this paper. In particular, they show that for the von-Mises distribution the KL-gross error sensitivity coincides with the s.g.e.s. of Ko and Guttorp (1988) for the choice of the dispersion measure $\left(\sqrt{\kappa A(\kappa)}\right)^{-1}$. With this dispersion measure, their work shows that for the family $\Im = \{vM(\mu, \kappa), \kappa > 0\}$ both the circular median and symmetrically 100α % trimmed circular mean are SB-robust for any $0 < \alpha < 1$.

For circular data, it has been argued that breakdown occurs when contamination causes the direction to change by 180°. Davies and Gather (2006) proposes an alternative concept called the 'definability breakdown point' which is based on the notion of definability of the estimator and not on bias. For rotation equivariant functionals a high definability breakdown point (close to the maximum of 0.5) is attainable at concentrated distributions but only low values of this measure is possible for more dispersed distributions.

Control charts introduced by Shewhart (1931) play a very important role in the control of manufacturing processes. It has gained increasing popularity over the past eight decades as an effective tool for detecting shifts in the process mean from its target value μ_0 and for detecting uncontrolled variation in the process. In recent years, the application of robust statistics in statistical process control has received more attention in research. Albers et al. (2004, 2006), Chan et al. (1988), David (1989), Rocke (1989, 1992), Stoumbos and Reynolds (2000), Vommi and Seetala (2005, 2007), Celano (2009), Chenouri et al. (2009), Hamid et al. (2009) has examined different aspects of robustness in control chart construction in univariate and multivariate set-ups. Some of the most widely used performance measures for control charts are False Alarm Probability (FAP), No Signal Probability (NSP), Average Sample Number when the process is in-control (ASN_0) and Average Sample Number when the process is out-of-control (ASN_1). Two of the performance measures FAP and NSP are bounded between 0 and 1. Thus, it is natural to investigate whether the estimators of these performance measures are SB-robust. The other two performance measures ASN_0 and ASN_1 can take values between 0 and ∞ but their robustness properties have not been studied in the literature to the best of our knowledge. Hence, we study robustness and SB-robustness of all the four performance measures for the Shewhart control chart for monitoring the mean of a process. We assume that the quality characteristic follows a normal distribution and that the rational subgroup size is one.

Suppose that the distributions F and G of the random variables X and Y are symmetric about μ and ν, respectively. Bickel and Lehmann (1976) define measures of dispersion as non-negative functionals $\tau(F)$ (also denoted as $\tau(X)$ where X is a random variable having distribution F) which satisfy the following conditions:

1. $\tau(kX) = |k|\tau(X)$
2. $\tau(X + b) = \tau(X)$ for all b
3. $\tau(F) \leq \tau(G)$ whenever G is more dispersed than F.

If the random variable X has a symmetric distribution then $\tau(X) = \tau(-X)$ and (1) above holds for all $k \neq 0$. An important class of dispersion measures for symmetric distributions F is provided by the functional $\tau(F) = \left(\int_0^1 (F_*^{-1}(t))^\gamma d\Lambda(t)\right)^{1/\gamma}$ where F is assumed to be symmetric about μ, F_* denotes the distribution of $|X - \mu|$, Λ is any probability distribution on (0, 1) and γ is any positive number. The standard deviation (SD) of F which is possibly the most widely used measure of dispersion for real-valued random variables is a special case of the above ($\gamma = 2$ and Λ the uniform distribution on (0, 1)). The random variable Y is said to be more dispersed about ν than the random variable X about μ if $|Y - \nu|$ is stochastically larger than $|X - \mu|$.

The structure of the paper is as follows. Section 2 describes some of the well known circular distributions. In Sect. 3, some of the available results on SB-robustness of circular mean and trimmed estimators of mean and concentration parameters are discussed. In Sect. 4, we discuss the SB-robustness of FAP and ASN_0 of control charts and Sect. 5 deals with the SB-robustness of NSP and ASN_1. Section 6 concludes the paper.

2 Circular Distributions

A circular random variable is a measurable map $\Theta : \Omega \longrightarrow T$ from a probability space Ω to the unit circle T. The distribution of Θ can be characterised by specifying the probabilities $P(\Theta \in [\alpha_1, \alpha_2))$ for all arcs $[\alpha_1, \alpha_2)$ traversed in anti-clockwise direction. The probability distribution of a circular random variable Θ has the property that $P(\Theta \in T) = 1$ and are often referred to as circular distribution. The probability density function (p.d.f) $f(\theta)$ of a circular random variable has the following properties:

1. $f(\theta) \geq 0$ for all $0 \leq \theta < 2\pi$
2. $\int_0^{2\pi} f(\theta)d\theta = 1$
3. $f(\theta) = f(\theta + 2\pi k)$ for any integer k and all $\theta, 0 \leq \theta < 2\pi$ (i.e., f is periodic).

The popular circular distributions include the von-Mises distribution (vM), the Wrapped Normal distribution (WN), the Wrapped Cauchy (WC) distribution, the Circular Uniform distribution(CU), the Cardiod distribution etc. A comprehensive account of the properties of these distributions can be found in Mardia and Jupp (2000), Jammalamadaka and SenGupta (2001). Among these, the von-Mises (a.k.a. Circular Normal) distribution is the most popular circular distribution for applied work. A circular random variable Θ is said to have a vM distribution with mean direction parameter μ and concentration parameter κ if it has the probability density function (p.d.f)

$$f(\theta; \mu, \kappa) = \frac{1}{2\pi I_0(\kappa)} \exp(\kappa \cos(\theta - \mu)), 0 \leq \theta < 2\pi, 0 \leq \mu < 2\pi, \kappa > 0$$

where $I_0(.)$ is the modified Bessel function of order 0. This distribution is symmetric about μ and is unimodal. We will denote this distribution as $vM(\mu, \kappa)$. For sufficiently large κ the $vM(\mu, \kappa)$ distribution can be approximated by a normal distribution with mean μ and variance $\frac{1}{\sqrt{\kappa}}$.

The maximum likelihood estimate (MLE) of the parameters μ and κ are given by $\hat{\mu} = arctan^*(\frac{S}{C})$ and $\hat{\kappa} = A^{-1}(\bar{R})$ where $\bar{R} = \frac{R}{n}$. The function $A(\kappa) = \frac{I_1(\kappa)}{I_0(\kappa)}$ has many interesting properties like

1. $0 \leq A(\kappa) < 1$
2. $A(\kappa) \longrightarrow 0$ as $\kappa \longrightarrow 0$
3. $A'(\kappa) = \frac{d}{d\kappa}[A(\kappa)] = [1 - \frac{A(\kappa)}{\kappa} - A^2(\kappa)] \geq 0$

i.e., $A(\kappa)$ is a strictly increasing function of κ. The MLE of μ does not depend on the value of κ but that is not the case for the MLE of κ (see Jammalamadaka and SenGupta (2001), pp. 86–88).

Another useful distribution on the circle is the Wrapped Normal (WN) distribution which is obtained by wrapping a $N(\mu, \sigma^2)$ distribution around the circle. Its probability density function is given by

$$g(\theta) = \frac{1}{\sigma\sqrt{2\pi}} \sum_{m=-\infty}^{\infty} exp[\frac{-(\theta - \mu - 2\pi m)^2}{2\sigma^2}]$$

Alternatively this density can be represented as

$$g(\theta) = \frac{1}{2\pi}(1 + 2\sum_{p=1}^{\infty} \rho^{p^2} cos\, p(\theta - \mu)), 0 \leq \theta < 2\pi, 0 \leq \mu < 2\pi, 0 < \rho < 1$$

Here, μ is the mean direction parameter and ρ is the concentration parameter. The parameters ρ and σ are related by $\rho = exp(\frac{-\sigma^2}{2})$. Like the vM distribution, the WN distribution is also unimodal and symmetric about the value $\theta = \mu$.

3 SB-Robustness of Circular Mean

Laha and Mahesh (2011), Laha et al. (2013) studied the robustness of circular mean for different families of circular distributions. They showed that the circular mean is SB-robust for the following families: (1) mixture of two von-Mises distributions (2) mixture of wrapped normal and von-Mises distributions, and (3) mixture of two wrapped normal distributions when the measure of dispersion is $S(F) = E_F(d(\Theta, \mu))$. As mentioned in Sect. 1, an estimator T may be SB-robust at the family of distributions $\Im$ for one choice of dispersion measure while it may not be so for another choice of dispersion measure. For example, He and Simpson (1992) showed that the circular mean is SB-robust estimator of μ at $\Im = \{vM(\mu, \kappa), \kappa > 0\}$ for the dispersion measure $\left(\sqrt{\kappa A(\kappa)}\right)^{-1}$ whereas Ko and

Guttorp (1988) showed that the circular mean is not SB-robust at $\Im$ when the dispersion measure is $\sqrt{1 - A(\kappa)}$. Laha and Mahesh (2011) proved that the circular mean is SB-robust at $F_1 = \{vM(\mu, \kappa), \kappa > m > 0\}$ for all the three dispersion measures mentioned above.

3.1 Equivalent Dispersion Measures

Suppose S_1 and S_2 are two dispersion measures defined on the family of distributions $\Im$. Then, S_1 and S_2 are said to be equivalent measures of dispersion for the family of distributions $\Im$ if $sup_{\Im} R(F)$ and $sup_{\Im} R^{-1}(F)$ are both finite, where $R(F) = \frac{S_1(F)}{S_2(F)}$. Laha and Mahesh (2011) introduced this concept and proved that if S_1 and S_2 are two equivalent measures of dispersion for the family of distributions $\Im$ and if the estimating functional T is SB-robust at $\Im$ when the measure of dispersion is S_2, then T is also SB-robust at $\Im$ when the measure of dispersion is S_1. Further they proved that $S_1 = \sqrt{1 - A(\kappa)}$, $S_2 = E_F(d(\Theta, 0))$, and $S_3 = (\kappa A(\kappa))^{-(1/2)}$ are equivalent measures of dispersion for the family of distributions $\Im^* = \{vM(0, \kappa); \kappa > m > 0\}$. But for the family $\Im = \{vM(\mu, \kappa); \kappa > 0\}$, they proved that S_1 and S_2 are equivalent measures of dispersion, but S_1 and S_3 are not, thus explaining the apparently conflicting results obtained in He and Simpson (1992), Ko and Guttorp (1988).

Laha et al. (2013) considered another family of distributions $\Im_1{}^* = \{WN(0, \rho); 0 < m < \rho < 1\}$. In this case they took four dispersion measures $S_1(F) = \sqrt{1 - \rho_F}$, $S_2(F) = E_F(d(\Theta, 0))$, $S_3(F) = (\rho_F A^{-1}(\rho_F))^{-(1/2)}$ and $S_4(F) = E_{\gamma,F}(d(\Theta, 0))$ (See Sect. 3.2 below for definition of $E_{\gamma,F}(d(\Theta, 0))$). They proved that all the above are equivalent measures of dispersion for the family $\Im_1{}^*$. However for the family of distributions $\tilde{\Im} = \{WN(0, \rho); 0 < \rho < 1\}$, S_1, S_2 and S_4 are equivalent measures of dispersion but S_2 and S_3 are not. The circular mean direction $T(F)$ is proved to be SB-robust for $\Im^{**} = \{WN(0, \rho); 0 < m < \rho < M < 1\}$ when the measure of dispersion is S_1. Using the concept of equivalence of dispersion measures and by noting that $\Im^{**} \subset \Im_1{}^*$, Laha et al. (2013) proved that it is also SB-robust at $\Im^{**}$ w.r.t. the dispersion measures S_2, S_3 and S_4. Also they show that $T(F)$ is not SB-robust for the family of distributions $\tilde{\Im}_1 = \{WN(\mu, \rho); 0 < \rho < 1\}$ when the measure of dispersion is $S_2(F) = E_F(d(\Theta, \mu))$ where $F \in \tilde{\Im}_1$. Hence, it is also not SB-robust for $\tilde{\Im}_1$ w.r.t.the equivalent dispersion measures S_1 and S_4.

3.2 SB-Robustness of Trimmed Estimators of Mean and Concentration Parameters

As the circular mean is B-robust but not SB-robust for $\Im = \{vM(\mu, \kappa), \kappa > 0\}$, Laha and Mahesh (2011) introduced the concept of γ-circular trimmed mean. Suppose Θ

is a circular random variable with p.d.f. $f(\theta)$ and $0 \le \gamma < 0.5$ is fixed. Let α, β be two points on the unit circle satisfying

1. $\int_{\beta}^{\alpha} f(\theta)d\theta = 1 - 2\gamma$
2. $d_1(\alpha, \beta) \le d_1(\mu, \nu)$ for all μ, ν satisfying $\int_{\nu}^{\mu} f(\theta)d\theta = 1 - 2\gamma$ where $d_1(\phi, \xi)$ is the length of the arc starting from ξ and ending at ϕ traversed in the anti-clockwise direction.

The γ-circular trimmed mean (γ-CTM) is then defined as

$$\mu_\gamma = arg[\frac{1}{1-2\gamma}\int_{\beta}^{\alpha} exp(i\theta)f(\theta)d\theta]$$

where γ is the trimming proportion. Laha and Mahesh (2011) proved that the γ-CTM (μ_γ) is SB-robust at the family of distributions $\Im = \{vM(\mu, \kappa), \kappa > 0\}$ when the measure of dispersion is $S(F) = E_{\gamma,F}(d(\Theta, \mu))$ where $F \in \Im$ and $0 \le \gamma < 0.5$

Ko and Guttorp (1988) proved that $K(F) = A^{-1}(\rho_F)$ where

$$\rho_F = \sqrt{E_F^2(cos\Theta) + E_F^2(sin\Theta)} = A(\kappa)$$

is not SB-robust at the family $\Im = \{vM(\mu, \kappa), \kappa > 0\}$ when the measure of dispersion is $S(F) = \sqrt{1 - A(\kappa)}$. Laha and Mahesh (2012) discussed robust estimation of κ for vM distribution. They showed that $K(F)$ is not SB-robust w.r.t. the dispersion measure $S(F) = E_F(d(\Theta, \mu))$. They proposed a new trimmed estimator for κ which is defined as follows: Let $f(\theta; \mu, \kappa)$ be the p.d.f. of $vM(\mu, \kappa)$ distribution and $\alpha(\kappa)$ and $\beta(\kappa)$ be symmetrically placed around μ such that

$$\int_{\beta(\kappa)}^{\alpha(\kappa)} f(\theta; \mu, \kappa)d\theta = 1 - 2\gamma$$

where γ is the trimming proportion such that $\gamma \in [0, 0.5)$. Define

$$g^*(\kappa) = E_{\gamma,F}(d(\Theta, \mu)) = \int_{\beta(\kappa)}^{\alpha(\kappa)} d(\theta, \mu)f(\theta; \mu, \kappa)d\theta$$

Then the new trimmed estimator for κ is defined as

$$T_\gamma(F) = g^{*-1}[E_{\gamma,F}(d(\Theta, \mu))]$$

Laha and Mahesh (2012) proved that if $\Theta \sim vM(0, \kappa)$, $d(\theta) = \pi - |\pi - \theta|$ and $g^*(\kappa) = E_{\gamma,F}(d(\Theta))$, then $T_\gamma(F) = g^{*-1}[E_{\gamma,F}(d(\Theta))]$ is SB-robust at the family of distributions $\Im^* = \{vM(0, \kappa); 0 < m \le \kappa \le M\}$ w.r.t. the dispersion measure $S(F) = E_{\gamma,F}(d(\Theta)) = (1 - 2\gamma)^{-1} \int_{\beta(\kappa)}^{\alpha(\kappa)} d(\theta)dF$. Similar results on SB-robustness of mean and concentration parameter of Wrapped Normal distribution can be seen in Laha et al. (2013).

4 SB-Robustness of *FAP* and ASN_0 of Control Chart

In the construction of control charts, one of the most important consideration is that of the False Alarm Probability (FAP). This is the probability that an observation falls outside the control limits of a control chart when the process distribution has not changed. The FAP is controlled at some chosen level α which is conventionally taken to be 0.0027. Suppose that a control chart has been constructed with $N(\mu_0, \sigma_0)$ as the underlying distribution of the quality characteristic and rational subgroup size 1. Then, the Lower Control Limit (*LCL*) and Upper Control Limit (*UCL*) of the $\bar{X}$ chart are $\mu_0 - 3\sigma_0$ and $\mu_0 + 3\sigma_0$, respectively. The FAP can be represented as a functional given below.

$T(F) = E_F(I_{\complement(LCL,UCL)}(X)) = P_F(X \leq LCL \text{ or } X \geq UCL)$ where

$$I_{\complement(LCL,UCL)}(X) = \begin{cases} 1 \text{ if } X \leq LCL \text{ or } X \geq UCL \\ 0 \qquad \text{otherwise.} \end{cases} \tag{1}$$

Here, $\complement(LCL, UCL)$ denotes the complement of the interval (LCL,UCL). When $F = N(\mu_0, \sigma_0)$ we have

$$T(F) = 1 - [\Phi(\frac{UCL - \mu_0}{\sigma_0}) - \Phi(\frac{LCL - \mu_0}{\sigma_0})] = \alpha \tag{2}$$

Let $\Im = \{N(\mu, \sigma_0) : k_1 \leq \mu \leq k_2\}$, $\mu_0 \in [k_1, k_2]$. In the theorem below we discuss the SB-robustness of FAP at the family $\Im$.

Theorem 1 *1. $T(F) = E_F(I_{\complement(LCL,UCL)}(X))$ is B-robust.*
2. $T(F)$ is SB-robust at the family $\Im$ when the measure of dispersion is $S(F) = \sqrt{T(F)(1 - T(F))}$.

Proof Let $G_\varepsilon = (1 - \varepsilon)F + \varepsilon\delta_x$. Then the functional corresponding to G_ε can be written as

$$\begin{aligned} T(G_\varepsilon) &= E_{(1-\varepsilon)F+\varepsilon\delta_x}(I_{\complement(LCL,UCL)}(X)) \\ &= (1 - \varepsilon)E_F(I_{\complement(LCL,UCL)}(X)) + \varepsilon(I_{\complement(LCL,UCL)}(x)) \\ &= (1 - \varepsilon)T(F) + \varepsilon \qquad \text{if } x \in \complement(LCL, UCL) \\ &= (1 - \varepsilon)T(F) \qquad \text{if } x \in (LCL, UCL) \end{aligned}$$

As defined earlier

$$IF(x; T, F) = \lim_{\varepsilon \to 0} \frac{T(G_\varepsilon) - T(F)}{\varepsilon}$$

Substituting $T(G_\varepsilon)$ in the above we get,

$$IF(x;T,F) = \begin{cases} \lim_{\varepsilon\to 0} \frac{[(1-\varepsilon)T(F)]+\varepsilon-T(F)}{\varepsilon} & \text{if } x \in \complement(LCL, UCL) \\ \lim_{\varepsilon\to 0} \frac{[(1-\varepsilon)T(F)]-T(F)}{\varepsilon} & \text{if } x \in (LCL, UCL) \end{cases}$$

A simple calculation yields

$$IF(x;T,F) = \begin{cases} 1 - T(F) & \text{if } x \in \complement(LCL, UCL) \\ -T(F) & \text{if } x \in (LCL, UCL) \end{cases} \tag{3}$$

Therefore, $\gamma(T,F) = sup_x|IF(x;T,F)| < \infty$ as $|IF(x;T,F)| < 1, \forall$x. Hence $T(F)$ is B-robust.
Now, the standardised influence function of T at F is

$$SIF(x;T,F,S) = \begin{cases} \sqrt{\frac{1-T(F)}{T(F)}} & \text{if } x \in \complement(LCL, UCL) \\ -\sqrt{\frac{T(F)}{1-T(F)}} & \text{if } x \in (LCL, UCL) \end{cases} \tag{4}$$

Now, when $F = N(\mu, \sigma_0)$, let $t_1 = \frac{\mu_0-\mu+3\sigma_0}{\sigma_0}$ and $t_2 = \frac{\mu_0-\mu-3\sigma_0}{\sigma_0}$. Then, we can write $T(F) = 1 - [\Phi(t_1) - \Phi(t_2)]$. Now it can be seen using simple calculus that $min_{\Im} T(F) > 0$ and the minimum is attained at $\mu = \mu_0$. Also we note the following:

1. $T(F) = h(\mu) = 1 - [\Phi(t_1) - \Phi(t_2)]$
2. h is a decreasing function of μ in the interval $(-\infty, \mu_0]$ and is an increasing function of μ in the interval $[\mu_0, \infty)$. Further $\lim_{\mu\to\infty} h(\mu) = 1$ and $\lim_{\mu\to-\infty} h(\mu) = 1$.

Thus we conclude that $max_{\Im} T(F) < 1$ and hence $\gamma^*(T, \Im, S) < \infty$. Hence the theorem. □

Remark 1 The FAP is not SB-robust at the family $\Im^* = \{N(\mu, \sigma_0) : -\infty < \mu < \infty\}$ w.r.t. the dispersion measure $S(F)$.

Theorem 2 *$T(F) = E_F(I_{\complement(LCL,UCL)}(X))$ is not SB-robust at the family $\Im_1$ when the measure of dispersion is $S(F) = \sqrt{T(F)(1-T(F))}$ where $\Im_1 = \{N(\mu_0, \sigma) : \sigma > 0\}$.*

Proof From Eq. 4 above, we have

$$SIF(x;T,F,S) = \begin{cases} \sqrt{\frac{1-T(F)}{T(F)}} & \text{if } x \in \complement(LCL, UCL) \\ -\sqrt{\frac{T(F)}{1-T(F)}} & \text{if } x \in (LCL, UCL) \end{cases}$$

Note that when $F = N(\mu_0, \sigma)$, we have $T(F) = g(\sigma) = 1 - [\Phi(t) - \Phi(-t)]$ where $t = \frac{3\sigma_0}{\sigma}$. Also we observe that as $\sigma \longrightarrow 0$, $g(\sigma) \longrightarrow 0$ and as $\sigma \longrightarrow \infty$, $g(\sigma) \longrightarrow 1$. Therefore, $\gamma^*(T, \Im_1, S) = \infty$. Hence the theorem. □

Remark 2 The FAP is SB-robust at the family $\Im_1^* = \{N(\mu_0, \sigma) : 0 < m_1 \leq \sigma \leq m_2\}$ w.r.t. the dispersion measure $S(F)$.

In Theorem 3 below we discuss the SB-robustness of the FAP at the family $\Im_2 = \{N(\mu, \sigma) : k_1 \leq \mu \leq k_2, 0 < m_1 \leq \sigma \leq m_2\}$, $\mu_0 \in [k_1, k_2]$.

Theorem 3 *$T(F) = E_F(I_{\complement(LCL,UCL)}(X))$ is SB-robust at the family $\Im_2$ when the measure of dispersion is $S(F) = \sqrt{T(F)(1 - T(F))}$.*

Proof From Eq. 4 above, we have

$$SIF(x; T, F, S) = \begin{cases} \sqrt{\frac{1-T(F)}{T(F)}} & \text{if } x \in \complement(LCL, UCL) \\ -\sqrt{\frac{T(F)}{1-T(F)}} & \text{if } x \in (LCL, UCL) \end{cases}$$

Note that when $F = N(\mu, \sigma)$, we can write $T(F) = 1 - [\Phi(t_3) - \Phi(t_4)]$ where $t_3 = \frac{\mu_0 - \mu + 3\sigma_0}{\sigma}$ and $t_4 = \frac{\mu_0 - \mu - 3\sigma_0}{\sigma}$. It can be seen using simple calculus that $min_{\Im_2}$ $T(F) = 1 - [\Phi(t_5) - \Phi(-t_5)] > 0$ and $max_{\Im_2} T(F) = 1 - [\Phi(t_6) - \Phi(-t_6)] < 1$ where $t_5 = \frac{3\sigma_0}{m_1}$ and $t_6 = \frac{3\sigma_0}{m_2}$. Thus $\gamma^*(T, \Im_2, S) < \infty$. Hence the theorem. □

Another important consideration for control chart performance is ASN_0, which is the average run length before an out-of-control signal is given by the control chart. It is expected that the ASN_0 should be large when the process is in-control and it should be small if the process is out-of-control. The ASN_0 can be represented as a functional given below.

$$T_1(F) = [E_F(I_{\complement(LCL,UCL)}(X))]^{-1} = \frac{1}{T(F)}$$

In Theorem 4 below we discuss the SB-robustness of ASN_0 at the family $\Im = \{N(\mu, \sigma_0) : k_1 \leq \mu \leq k_2\}$, $\mu_0 \in [k_1, k_2]$.

Theorem 4 *1. $T_1(F) = [E_F(I_{\complement(LCL,UCL)}(X))]^{-1} = \frac{1}{T(F)}$ is B-robust.*
2. $T_1(F)$ is SB-robust at the family $\Im$ with respect to the dispersion measure $S^*(F) = \frac{1-T(F)}{(T(F))^2}$.

Proof Let $G_\varepsilon = (1 - \varepsilon)F + \varepsilon\delta_x$ where $F = N(\mu, \sigma_0)$. The functional corresponding to G_ε can be written as

$$T_1(G_\varepsilon) = \begin{cases} \frac{1}{(1-\varepsilon)E_F(I_{\complement(LCL,UCL)}(X))+\varepsilon} & \text{if } x \in \complement(LCL, UCL) \\ \frac{1}{(1-\varepsilon)E_F(I_{\complement(LCL,UCL)}(X))} & \text{if } x \in (LCL, UCL) \end{cases}$$

The influence function of T_1 at F is

$$IF(x; T_1, F) = \lim_{\varepsilon \to 0} \frac{T_1(G_\varepsilon) - T_1(F)}{\varepsilon}$$

Thus,

$$IF(x; T_1, F) = \begin{cases} \lim_{\varepsilon \to 0} \frac{\frac{1}{(1-\varepsilon)T(F)+\varepsilon} - \frac{1}{T(F)}}{\varepsilon} & \text{if } x \in \complement(LCL, UCL) \\ \lim_{\varepsilon \to 0} \frac{\frac{1}{(1-\varepsilon)T(F)} - \frac{1}{T(F)}}{\varepsilon} & \text{if } x \in (LCL, UCL) \end{cases}$$

which on simplification and applying the limits gives

$$IF(x; T_1, F) = \begin{cases} \frac{T(F)-1}{T(F)^2} \text{ if } x \in \complement(LCL, UCL) \\ \frac{1}{T(F)} \quad \text{if } x \in (LCL, UCL) \end{cases} \tag{5}$$

Hence $\sup_x |IF(x; T_1, F)| < \infty$. Thus, T_1 is B-robust.

Now to check the SB-robustness of T_1, we note that

$$SIF(x, T_1, F, S^*) = \begin{cases} -1 & \text{if } x \in \complement(LCL, UCL) \\ \frac{T(F)}{1-T(F)} & \text{if } x \in (LCL, UCL) \end{cases} \tag{6}$$

As noted in Theorem 1, $max_{\Im} T(F) < 1$. Thus $\gamma^*(T_1, \Im, S^*) < \infty$. Hence the theorem. □

Remark 3 The ASN_0 is not SB-robust at the family $\Im^* = \{N(\mu, \sigma_0) : -\infty < \mu < \infty\}$ w.r.t. the dispersion measure $S^*(F)$.

In Theorem 5 below, we discuss the SB-robustness of ASN_0 at the family $\Im_2 = \{N(\mu, \sigma) : k_1 \leq \mu \leq k_2, 0 < m_1 \leq \sigma \leq m_2\}$, $\mu_0 \in [k_1, k_2]$.

Theorem 5 $T_1(F) = [E_F(I_{\complement(LCL,UCL)}(X))]^{-1} = \frac{1}{T(F)}$ *is SB-robust at the family* $\Im_2$ *when the measure of dispersion is* $S^*(F) = \frac{1-T(F)}{(T(F))^2}$.

Proof From Eq. 6 above,

$$SIF(x, T_1, F, S^*) = \begin{cases} -1 & \text{if } x \in \complement(LCL, UCL) \\ \frac{T(F)}{1-T(F)}; & \text{if } x \in (LCL, UCL) \end{cases}$$

Now note that when $F = N(\mu, \sigma)$, $T(F) = 1 - [\Phi(t_3) - \Phi(t_4)]$. It can be proved using arguments similar to that given in the proof of Theorem 3 that $\gamma^*(T_1, \Im_2, S^*) < \infty$. Hence the theorem. □

5 SB-Robustness of *NSP* and ASN_1

It is expected that the control chart would be able to detect quickly that a process is out-of-control. In terms of measures of performance one usually considers the NSP, which is defined as the P (out-of-control signal is not given by the control chart when the process is out-of-control), and the ASN_1 (which is the ASN when the process is out-of-control). It is expected that NSP and ASN_1 be both small when the process is out-of-control. In this section, we study the SB-robustness of these two measures NSP and ASN_1. In the Theorem 6 below, we discuss the SB-robustness of the NSP when the process is out-of-control at the family $\Im = \{N(\mu, \sigma_0) : k_1 \leq \mu \leq k_2\}$, $\mu_0 \in [k_1, k_2]$.

Theorem 6 *1. $T_2(F) = E_F[I_{(LCL,UCL)}(X)]$ is B-robust.*
2. $T_2(F)$ is SB-robust at the family $\Im$ when the measure of dispersion is $S^{**}(F) = \sqrt{T_2(F)(1 - T_2(F))}$.

Proof Let $F = N(\mu, \sigma_0)$, $T_2(F) = \beta(\mu)$. Then we have, $\beta(\mu) = P(t_2 < \frac{X-\mu}{\sigma_0} < t_1) = \Phi(t_1) - \Phi(t_2)$, where t_1 and t_2 are defined in Theorem 1. Let $G_\varepsilon = (1 - \varepsilon)F + \varepsilon\delta_x$. Then the functional corresponding to G_ε can be written as

$$T_2(G_\varepsilon) = E_{(1-\varepsilon)F+\varepsilon\delta_x}(I_{(LCL,UCL)})(X) = \begin{cases} (1-\varepsilon)T_2(F) + \varepsilon & \text{if } x \in (LCL, UCL) \\ (1-\varepsilon)T_2(F) & \text{if } x \notin (LCL, UCL) \end{cases}$$

Hence

$$IF(x; T_2, F) = \begin{cases} 1 - T_2(F) & \text{if } x \in (LCL, UCL) \\ -T_2(F) & \text{if } x \notin (LCL, UCL) \end{cases} \quad (7)$$

Thus $\gamma(T_2, F) < \infty$ as $|IF(x; T_2, F)| \leq 1, \forall x$. Hence $T_2(F)$ is B-robust.
To check the SB-robustness of T_2 at F, we consider

$$SIF(x, T_2, F, S^{**}) = \begin{cases} \sqrt{\frac{1-T_2(F)}{T_2(F)}} & \text{if } x \in (LCL, UCL) \\ -\sqrt{\frac{T_2(F)}{1-T_2(F)}} & \text{if } x \notin (LCL, UCL) \end{cases} \quad (8)$$

Now, note that when $F = N(\mu, \sigma_0)$, we have $T_2(F) = \beta(\mu) = [\Phi(t_1) - \Phi(t_2)]$. As earlier we can see using simple calculus that maximum value of $T_2(F)$ as F varies over $\Im$ is attained at $\mu = \mu_0$ and $max_{\Im} T_2(F) < 1$. We also note the following.

1. $\beta(\mu) = [\Phi(t_1) - \Phi(t_2)]$
2. β is increasing in the interval $(-\infty, \mu_0]$ and decreasing in the interval $[\mu_0, \infty)$, as $lim_{\mu\to\infty}\beta(\mu) = 0$ and $lim_{\mu\to-\infty}\beta(\mu) = 0$.

Therefore, we conclude that $min_{\Im} T_2(F) > 0$ and thus $\gamma^*(T_2, \Im, S^{**}) < \infty$. Hence the theorem. □

Remark 4 The *NSP* is not SB-robust at the family $\Im^* = \{N(\mu, \sigma_0) : -\infty < \mu < \infty\}$ w.r.t. the dispersion measure $S^{**}(F)$.
In the theorem below, we discuss the SB-robustness of the ASN_1 at the family $\Im$.

Theorem 7 *1. $T_3(F) = E_F[I_{(LCL,UCL)}(X)]^{-1} = \frac{1}{T_2(F)}$ is B-robust.*
2. $T_3(F)$ is SB-robust at the family $\Im$ when the measure of dispersion is $S^{***}(F) = \frac{1-T_2(F)}{[T_2(F)]^2}$.

Proof Let $G_\varepsilon = (1-\varepsilon)F + \varepsilon\delta_x$. Then, the functional corresponding to G_ε can be written as

$$T_3(G_\varepsilon) = \begin{cases} \frac{1}{(1-\varepsilon)E_F[I_{(LCL,UCL)}(X)]+\varepsilon} & \text{if } x \in (LCL, UCL); \\ \frac{1}{(1-\varepsilon)E_F[I_{(LCL,UCL)}(X)]} & \text{if } x \notin (LCL, UCL) \end{cases}$$

The influence function is

$$IF(x; T_3, F) = \lim_{\varepsilon \to 0} \frac{T_3(G_\varepsilon) - T_3(F)}{\varepsilon}$$

Thus

$$IF(x; T_3, F) = \begin{cases} \lim_{\varepsilon \to 0} \frac{\frac{1}{(1-\varepsilon)T_2(F)+\varepsilon} - \frac{1}{T_2(F)}}{\varepsilon} & \text{if } x \in (LCL, UCL) \\ \lim_{\varepsilon \to 0} \frac{\frac{1}{(1-\varepsilon)T_2(F)} - \frac{1}{T_2(F)}}{\varepsilon} & \text{if } x \notin (LCL, UCL) \end{cases} \tag{9}$$

On simplification we get,

$$IF(x; T_3, F) = \begin{cases} \frac{T_2(F)-1}{[T_2(F)]^2} & \text{if } x \in (LCL, UCL) \\ \frac{1}{T_2(F)} & \text{if } x \notin (LCL, UCL) \end{cases} \tag{10}$$

Therefore $\gamma(T_3, F) < \infty$ and hence T_3 is B-robust at F. Now to check the SB-robustness of T_3 at F, we consider

$$SIF(x; T_3, F, S^{***}) = \begin{cases} -1 & \text{if } x \in (LCL, UCL) \\ \frac{T_2(F)}{1-T_2(F)} & \text{if } x \notin (LCL, UCL) \end{cases} \tag{11}$$

Now, note that when $F = N(\mu, \sigma_0)$, we have $T_2(F) = [\Phi(t_1) - \Phi(t_2)]$. Arguing as in the proof of Theorem 6, we get $max_{\Im} T_2(F) < 1$ and $min_{\Im} T_2(F) > 0$ which implies that $\gamma^*(T_3, \Im, S^{***}) < \infty$. Hence the theorem. □

In the theorem below, we discuss the SB-robustness of the ASN_1 at the family $\Im_2 = \{N(\mu, \sigma) : k_1 \le \mu \le k_2, 0 < m_1 \le \sigma \le m_2\}$, $\mu_0 \in [k_1, k_2]$.

Theorem 8 $T_3(F) = E_F[I_{(LCL,UCL)}(X)]^{-1} = \frac{1}{T_2(F)}$ *is SB-robust at the family* $\Im_2$ *when the measure of dispersion is* $S^{***}(F) = \frac{1-T_2(F)}{[T_2(F)]^2}$.

Proof From Eq. 11 above

$$SIF(x; T_3, F, S^{***}) = \begin{cases} -1 & \text{if } x \in (LCL, UCL) \\ \frac{T_2(F)}{1-T_2(F)} & \text{if } x \notin (LCL, UCL) \end{cases}$$

Now, note that when $F = N(\mu, \sigma)$, we have $T_2(F) = [\Phi(t_3) - \Phi(t_4)]$. Arguing as in the proof of Theorem 7, we get $max_{\Im_2} T_2(F) < 1$ and $min_{\Im_2} T_2(F) > 0$ which implies that $\gamma^*(T_3, \Im_2, S^{***}) < \infty$. Hence the theorem. □

6 Conclusion

In this paper, we have reviewed the SB-robustness of estimators of mean and concentration parameters of von-Mises distribution. Laha and Mahesh (2011, 2012), Laha et al. (2013) discusses the shortcomings of the MLEs of circular mean direction and concentration parameter with respect to SB-robustness and proposes new estimators which have better properties. Considering the importance of robustness of performance measures in statistical process control, we have examined the SB-robustness of some of the control chart performance measures in this article. It is seen that the FAP, NSP, ASN_0 and ASN_1 are all B-robust. They are also SB-robust at the family $\Im_2 = \{N(\mu, \sigma) : k_1 \leq \mu \leq k_2, 0 < m_1 \leq \sigma \leq m_2\}$, $\mu_0 \in [k_1, k_2]$. However, none of these performance measures are SB-robust when we consider the larger family of distributions $\{N(\mu, \sigma) : -\infty < \mu < \infty, \sigma > 0\}$. This opens the possibility of considering alternative measures which are SB-robust for this larger family. We intend to take this up for studying in a future paper.

Acknowledgments The authors thank the referees of this paper and one of the editors of this volume for their helpful comments which have improved this paper.

References

Agostinelli C (2007) Robust estimation for circular data. Comput Stat Data Anal 51:5867–5875
Albers W, Kallenberg W, Nurdiati S (2004) Parametric control charts. J Stat Plann Infer 124:159–184
Albers W, Kallenberg W, Nurdiati S (2006) Data driven choice of control charts. J Stat Plann Infer 136:909–941
Bickel P, Lehmann E (1976) Descriptive statistics for nonparametric models. iii. dispersion. Ann Stat 4:1139–1158
Celano G (2009) Robust design of adaptive control charts for manual manufacturing/inspection workstations. J Appl Stat 36:181–203
Chan L, Hapuarachchi K, Macpherson B (1988) Robustness of $\bar{X}$ and r charts. IEEE Trans Reliab 37:117–123
Chenouri S, Steiner S, Variyath A (2009) A multivariate robust control chart for individual observations. J Qual Technol 41:259–271
David M (1989) Robust control charts. Technometrics 31:173–184
Davies L, Gather U (2006) Addendum to the discussion of "breakdown and groups". Ann Stat 3:1577–1579
Hamid S, Alireza M, Amir H (2009) A robust dispersion control chart based on m-estimate. J Ind Syst Eng 2:297–307
Hampel F (1968) Contribution to the theory of robust estimation. PhD thesis, University of California, Berkeley
Hampel F (1974) The influence curve and its role in robust estimation. J Am Stat Assoc 69:383–393
He X, Simpson D (1992) Robust direction estimation. Ann Stat 20:351–369
Huber P (1964) Robust estimation of a location parameter. Ann Math Stat 35:73–101
Jammalamadaka S, SenGupta A (2001) Topics in circular statistics. World Scientific, Singapore
Ko D, Guttorp P (1988) Robustness of estimators for directional data. Ann Stat 16:609–618

Laha AK, Mahesh KC (2011) SB-robustness of directional mean for circular distributions. J Stat Plann Infer 141:1269–1276

Laha AK, Mahesh KC (2012) SB-robust estimator for the concentration parameter of circular normal distribution. Stat Papers 53:457–467

Laha AK, Mahesh KC, Ghosh DK (2013) SB-robust estimators of the parameters of the wrapped normal distribution. Commun Stat Theor Meth 42:660–672

Mardia K, Jupp P (2000) Directional statistics. Wiley, Chichester

Rocke DM (1989) Robust control charts. Technometrics 31:173–184

Rocke DM (1992) $\bar{X}$ and r charts: robust control chart. The Statistician 41:97–104

Rousseeuw P (1981) A new infinitesimal approach to robust estimation. Zeitschrift fuer Wahrscheinlichkeit und Verwandte Gebiete 56:127–132

Shewhart W (1931) Economic control of quality of manufactured product. D.Van Nostrand Co., New York

Stoumbos Z, Reynolds M (2000) Robustness to non-normality and autocorrelation of individuals control charts. J Stat Comput Simul 66:145–187

Vommi V, Seetala M (2005) Applied soft computing. Technometrics 7:211–228

Vommi V, Seetala M (2007) A simple approach for robust economic design of control charts. Comput Oper Res 34:2001–2009

Wehrly T, Shine E (1981) Influence curves of estimators for directional data. Biometrika 68:334–335

Combining Linear Dimension Reduction Subspaces

Eero Liski, Klaus Nordhausen, Hannu Oja and Anne Ruiz-Gazen

1 Introduction

Dimension reduction plays an important role in high dimensional data analysis. In dimension reduction one wishes to reduce the dimension of a p-variate random vector $\mathbf{x} = (x_1, \ldots, x_p)^t$ using a transformation $\mathbf{z} = \mathbf{B}^t\mathbf{x}$, where the transformation matrix $\mathbf{B}$ is a $p \times k$ matrix with linearly independent columns, $k \leq p$. The column vectors of $\mathbf{B}$ then span the k-dimensional subspace of interest. The transformation to the subspace can also be done using the corresponding $p \times p$ orthogonal projection $\mathbf{P_B} = \mathbf{B}(\mathbf{B}^t\mathbf{B})^{-1}\mathbf{B}^t$. The transformation $\mathbf{z} = \mathbf{P_B}\mathbf{x}$ projects the observations onto a linear k-variate subspace.

There are two major types of dimension reduction methods, supervised and unsupervised dimension reduction. Unsupervised methods such as principal component analysis (PCA) or independent component analysis (ICA) reduce the dimension of $\mathbf{x}$ by trying to lose as little information as possible; in PCA the information loss is measured in terms of variance and in ICA in terms of non-gaussianity. In supervised dimension reduction, the goal is to reduce the dimension of $\mathbf{x}$ without losing any information on the dependence between $\mathbf{x}$ and a response variable y. It is then hoped that y is independent from $\mathbf{x}$ conditionally on $\mathbf{B}^t\mathbf{x}$. Popular supervised dimension

E. Liski (✉) · K. Nordhausen
University of Tampere, 33014 Tampere, Finland
e-mail: liskieero@gmail.com

K. Nordhausen
e-mail: klaus.nordhausen@utu.fi

K. Nordhausen · H. Oja
University of Turku, 20014 Turku, Finland
e-mail: hannu.oja@utu.fi

A. Ruiz-Gazen
Toulouse School of Economics, 31000 Toulouse, France
e-mail: anne.ruiz-gazen@tse-fr.eu

C. Agostinelli et al. (eds.), *Recent Advances in Robust Statistics: Theory and Applications*, DOI 10.1007/978-81-322-3643-6_7

reduction methods are for example SIR, SAVE, PHD and SICS (see Li 1991; Cook and Weisberg 1991; Li 1992; Liski et al. 2014a, respectively).

For practical problems, several dimension reduction methods are often available and it is difficult to decide which one to use. Individual dimension reduction methods are usually adequate to find only special types of subspaces or special types of relationships between $\mathbf{x}$ and y. In statistical learning, it is a current state of the art not to rely on a single learner but to train simultaneously many learners and combine them instead. Such combination rules are often called ensemble methods. See e.g. Zhou (2012) for a nice overview with explanations why ensemble methods are beneficial in practice. The goal in this paper is to use an ensemble of several dimension reduction methods. To combine the different dimension reduction methods is to say that we combine the individual orthogonal projections possibly with various ranks and find an "average orthogonal projection" (AOP) with an optimized rank. Our approach is similar to the approach in Crone and Crosby (1995). The idea is to find the AOP which is, on average, closest to the individual orthogonal projections with respect to some distance criterion.

The paper is organized as follows. In Sect. 2 we discuss subspaces and propose a generalization of the Crone and Crosby distance. Crone and Crosby (1995) considered subspaces of equal dimensions, whereas our weighted distance allows subspaces of different dimensions. Some natural choices of weights are given. Furthermore, the concept of averages of subspaces is discussed. In Sect. 3 the performance of the weighted distance and the different AOPs is evaluated in two unsupervised dimension reduction applications, one supervised dimension reduction simulation study and a real data example. Different dimension reduction methods may be complementary. The examples illustrate that the average of the associated orthogonal projections will make the most of them in the sense that (i) the AOP might outperform any individual dimension reduction method and (ii) the AOP is hardly affected by a few bad dimension reduction methods. In particular, when several methods coincide on some projection directions, the average orthogonal projection takes them into account and downplays projection directions rarely found. The paper ends with some final remarks.

2 Subspaces and Distances Between Subspaces

2.1 Subspaces with the Same Dimension k

We first consider linear subspaces in $\mathbb{R}^p$ with a fixed dimension k, $1 \leq k < p$. A linear subspace and the distances between subspaces can be defined in several ways.

1. The subspace is defined as a linear subspace spanned by the linearly independent columns of a $p \times k$ matrix $\mathbf{B}$, that is, $\mathscr{S}_{\mathbf{B}} = \{\mathbf{Ba} \ : \mathbf{a} \in \mathbb{R}^k\}$. This definition based on a matrix $\mathbf{B}$ is a bit obscure in the sense that $\mathscr{S}_{\mathbf{B}} = \mathscr{S}_{\mathbf{BA}}$ for all full-rank $k \times k$ matrices $\mathbf{A}$. According to this definition, the same subspace can in fact be fixed

by any member in a set of matrices equivalent to $\mathbf{B}$,

$$\{\mathbf{BA} \ : \ \mathbf{A} \text{ is a full-rank } k \times k \text{ matrix}\}.$$

The non-uniqueness of $\mathbf{B}$ may cause technical problems in the estimation of a subspace. Consider two $p \times k$ matrices $\mathbf{B}_1$ and $\mathbf{B}_2$ with rank k. Then a measure of distance between subspaces spanned by $\mathbf{B}_1$ and $\mathbf{B}_2$ can be defined as $k - \sum_{i=1}^k \rho_i^2 = k - tr(\mathbf{R}^t\mathbf{R})$ where $\rho_1^2, \ldots, \rho_k^2$ are the squared canonical correlations between $\mathbf{B}_1$ and $\mathbf{B}_2$ (Hotelling 1936) and

$$\mathbf{R} = (\mathbf{B}_1^t\mathbf{B}_1)^{-1/2}\mathbf{B}_1^t\mathbf{B}_2(\mathbf{B}_2^t\mathbf{B}_2)^{-1/2} .$$

Note that if $\mathbf{B}_1$ and $\mathbf{B}_2$ are equivalent then the squared canonical correlations are all 1.

2. The subspace is defined as a linear subspace spanned by the orthonormal columns of a $p \times k$ matrix $\mathbf{U}$. Note that, starting with $\mathbf{B}$, one can choose $\mathbf{U} = \mathbf{B}(\mathbf{B}^t\mathbf{B})^{-1/2}$ for this second definition. Unfortunately, the definition is still obscure as $\mathcal{S}_\mathbf{U} = \mathcal{S}_{\mathbf{UV}}$ for all orthonormal $k \times k$ matrices $\mathbf{V}$, and the same subspace is given by any matrix in the class of equivalent orthonormal matrices

$$\{\mathbf{UV} \ : \ \mathbf{V} \text{ is an orthonormal } k \times k \text{ matrix}\}.$$

The principal angles $\theta_i \in [0, \pi/2]$ between the subspaces $\mathbf{U}_1$ and $\mathbf{U}_2$ with corresponding k-variate direction vectors $\mathbf{u}_i$ and $\mathbf{v}_i$ $i = 1, \ldots, k$, are recursively defined by maximizing $\mathbf{u}_i^t(\mathbf{U}_1^t\mathbf{U}_2)\mathbf{v}_i$ subject to the constraints $\mathbf{u}_i^t\mathbf{u}_i = \mathbf{v}_i^t\mathbf{v}_i = 1$, and $\mathbf{u}_i^t\mathbf{u}_j = \mathbf{v}_i^t\mathbf{v}_j = 0$, $j = 1, \ldots, i-1$. The ith principal angle is then such that $\cos\theta_i = \mathbf{u}_i^t(\mathbf{U}_1^t\mathbf{U}_2)\mathbf{v}_i$, $i = 1, ..., k$, and a measure of distance between the subspaces may be obtained as $k - \sum_{i=1}^k \cos^2\theta_i = k - \sum_{i=1}^k (\mathbf{u}_i^t\mathbf{v}_i)^2$. It is easy to see that it equals to $k - \sum_{i=1}^k \rho_i^2$.

3. The subspace is defined as the linear subspace given by an orthogonal projection $\mathbf{P}$, that is, a $p \times p$ transformation matrix $\mathbf{P}$ such that

$$(\mathbf{x}_1 - \mathbf{P}\mathbf{x}_1) \perp \mathbf{P}\mathbf{x}_2 \text{ for all } \mathbf{x}_1, \mathbf{x}_2 \in \mathbb{R}^p \ \text{ which is equivalent to } \ \mathbf{P} = \mathbf{P}^t = \mathbf{P}^2.$$

The matrix $\mathbf{P}$ provides a unique way to fix the subspace $\mathcal{S}_\mathbf{P} = \{\mathbf{Px} : \mathbf{x} \in \mathbb{R}^p\}$. Note that, starting from $\mathbf{B}$, one can define $\mathbf{P} = \mathbf{P}_\mathbf{B} = \mathbf{B}(\mathbf{B}^t\mathbf{B})^{-1}\mathbf{B}^t$ in a unique way. Starting from $\mathbf{U}$ gives similarly $\mathbf{P} = \mathbf{P}_\mathbf{U} = \mathbf{UU}^t$. The squared distance between the subspaces given by two orthogonal projections $\mathbf{P}_1$ and $\mathbf{P}_2$ may then be defined as the matrix (Frobenius) norm

$$||\mathbf{P}_1 - \mathbf{P}_2||^2 = 2(k - tr(\mathbf{P}_1\mathbf{P}_2)) = 2(k - \sum_{i=1}^k \cos^2\theta_i) = 2(k - \sum_{i=1}^k \rho_i^2).$$

Crone and Crosby (1995) use

$$D(\mathbf{P}_1, \mathbf{P}_2) = (k - tr(\mathbf{P}_1\mathbf{P}_2))^{1/2} = \frac{1}{\sqrt{2}} \,||\mathbf{P}_1 - \mathbf{P}_2||$$

as a distance between two k-dimensional subspaces of $\mathbb{R}^p$ given by orthogonal projections $\mathbf{P}_1$ and $\mathbf{P}_2$. It is then easy to see that $0 \leq D^2(\mathbf{P}_1, \mathbf{P}_2) \leq min\{k, p - k\}$ and, because it is based on a norm, that the distance obeys the triangular inequality $D(\mathbf{P}_1, \mathbf{P}_3) \leq D(\mathbf{P}_1, \mathbf{P}_2) + D(\mathbf{P}_2, \mathbf{P}_3)$ for any orthogonal projections $\mathbf{P}_1, \mathbf{P}_2$, and $\mathbf{P}_3$.

2.2 Subspaces with Arbitrary Dimensions

Assume next that the ranks of the orthogonal projections $\mathbf{P}_1$ and $\mathbf{P}_2$ are k_1 and k_2, respectively, where $k_1, k_2 = 0, ..., p$. For completeness of the theory, we also accept projections $\mathbf{P} = \mathbf{0}$ with rank $k = 0$. As $||\mathbf{P}_1 - \mathbf{P}_2||^2 \geq |k_1 - k_2|$, one possible extension of the above distance is $D(\mathbf{P}_1, \mathbf{P}_2) = \frac{1}{\sqrt{2}} \left[||\mathbf{P}_1 - \mathbf{P}_2||^2 - |k_1 - k_2|\right]^{1/2}$. Then $0 \leq D^2(\mathbf{P}_1, \mathbf{P}_2) \leq min\{k_1, k_2, p - k_1, p - k_2\}$ but, unfortunately, the triangular inequality is not true for this distance. We therefore consider other extensions of the metric by Crone and Crosby (1995).

Let $w(0), \ldots, w(p)$ be positive weights attached to dimensions $0, \ldots, p$. (We will later see that the choice of $w(0)$ is irrelevant for the theory.) We then give the following definition.

Definition 1 A weighted distance between subspaces $\mathbf{P}_1$ and $\mathbf{P}_2$ with ranks k_1 and k_2 is given by

$$D_w^2(\mathbf{P}_1, \mathbf{P}_2) = \frac{1}{2} \,||w(k_1)\mathbf{P}_1 - w(k_2)\mathbf{P}_2||^2 . \tag{1}$$

The weights are used to make the orthogonal projections $\mathbf{P}_1$ and $\mathbf{P}_2$ with different ranks more comparable in some sense (see below some illustrating special cases). As the distance $D_w(\mathbf{P}_1, \mathbf{P}_2)$ is based on the matrix (Frobenius) norm, (i) $D_w(\mathbf{P}_1, \mathbf{P}_2) \geq 0$, (ii) $D_w(\mathbf{P}_1, \mathbf{P}_2) = 0$ if and only if $\mathbf{P}_1 = \mathbf{P}_2$, (iii) $D_w(\mathbf{P}_1, \mathbf{P}_2) = D_w(\mathbf{P}_2, \mathbf{P}_1)$, and (iv) $D_w(\mathbf{P}_1, \mathbf{P}_3) \leq D_w(\mathbf{P}_1, \mathbf{P}_2) + D_w(\mathbf{P}_2, \mathbf{P}_3)$, and we have the following results.

Lemma 1 *For two $p \times p$ orthogonal projections $\mathbf{P}_1$ and $\mathbf{P}_2$ with ranks k_1 and k_2, respectively*

$$\max\{p - k_1 - k_2, 0\} \leq tr(\mathbf{P}_1\mathbf{P}_2) \leq \min\{k_1, k_2\}.$$

Proof First note that $\mathbf{P}_1 = \mathbf{U}_1\mathbf{U}_1^t$ and $\mathbf{P}_2 = \mathbf{U}_2\mathbf{U}_2^t$ where $\mathbf{U}_1$ has k_1 orthonormal columns and $\mathbf{U}_2$ has k_2 orthonormal columns. Then $tr(\mathbf{P}_1\mathbf{P}_2) = ||\mathbf{U}_1^t\mathbf{U}_2||^2 \geq 0$. As $tr(\mathbf{P}_1\mathbf{P}_2) + tr(\mathbf{P}_1(\mathbf{I}_p - \mathbf{P}_2)) = tr(\mathbf{P}_1) = k_1$ and $tr(\mathbf{P}_1\mathbf{P}_2) + tr((\mathbf{I}_p - \mathbf{P}_1)\mathbf{P}_2) = tr(\mathbf{P}_2) = k_2$ one can conclude that $tr(\mathbf{P}_1\mathbf{P}_2) \leq \min\{k_1, k_2\}$. Similarly, $tr(\mathbf{P}_1(\mathbf{I}_p -$

$\mathbf{P}_2)) \leq \min\{k_1, p - k_2\}$ and therefore $tr(\mathbf{P}_1\mathbf{P}_2) = k_1 - tr(\mathbf{P}_1(\mathbf{I}_p - \mathbf{P}_2)) \geq k_1 - \min\{k_1, p - k_2\} = \max\{k_1 + k_2 - p, 0\}$, and the result follows. □

Note also that the lower and upper bounds in the above lemma are strict. The upper bound is obtained with the choices

$$\mathbf{P}_1 = \sum_{i=1}^{k_1} \mathbf{e}_i \mathbf{e}_i^t \text{ and } \mathbf{P}_2 = \sum_{i=1}^{k_2} \mathbf{e}_i \mathbf{e}_i^t,$$

and the lower bound with the choices

$$\mathbf{P}_1 = \sum_{i=1}^{k_1} \mathbf{e}_i \mathbf{e}_i^t \text{ and } \mathbf{P}_2 = \sum_{i=p-k_2+1}^{p} \mathbf{e}_i \mathbf{e}_i^t,$$

where $\mathbf{e}_i$ is a p-vector with the ith component one and other components zero.

Proposition 1 *For all weight functions w, $D_w(\mathbf{P}_1, \mathbf{P}_2)$ is a metric in the space of orthogonal projections, and the strict lower and upper bounds of $D_w^2(\mathbf{P}_1, \mathbf{P}_2)$ for the dimensions k_1 and k_2 are*

$$m(k_1, k_2) - w(k_1)w(k_2) \min\{k_1, k_2\} \leq D_w^2(\mathbf{P}_1, \mathbf{P}_2) \leq$$
$$m(k_1, k_2) + w(k_1)w(k_2) \min\{p - k_1 - k_2, 0\}$$

where

$$m(k_1, k_2) = \frac{w^2(k_1)k_1 + w^2(k_2)k_2}{2}.$$

Proof One easily sees that

$$\begin{aligned} D_w^2(\mathbf{P}_1, \mathbf{P}_2) &= \frac{w^2(k_1)k_1 + w^2(k_2)k_2}{2} - w(k_1)w(k_2)tr(\mathbf{P}_1\mathbf{P}_2) \\ &= m(k_1, k_2) - w(k_1)w(k_2)tr(\mathbf{P}_1\mathbf{P}_2), \end{aligned}$$

and the proof follows from Lemma 1. □

Some interesting choices of the weights are, for $k > 0$

$$(a)\ w_a(k) = 1, \quad (b)\ w_b(k) = \frac{1}{k}, \text{ and } (c)\ w_c(k) = \frac{1}{\sqrt{k}}.$$

Weights in (a) give the distance by Crone and Crosby (1995). Weights in (b) and (c) standardize the matrices so that $tr(w(k_i)\mathbf{P}_i) = 1$ and $||w(k_i)\mathbf{P}_i|| = 1$, respectively, if $k_i > 0$. It is remarkable that

$$D^2_{w_c}(\mathbf{P}_1, \mathbf{P}_2) = 1 - \frac{tr(\mathbf{P}_1\mathbf{P}_2)}{\sqrt{tr(\mathbf{P}_1)tr(\mathbf{P}_2)}}$$

where

$$\frac{tr(\mathbf{P}_1\mathbf{P}_2)}{\sqrt{tr(\mathbf{P}_1)tr(\mathbf{P}_2)}} = \frac{vec(\mathbf{P}_1)^t vec(\mathbf{P}_2)}{\sqrt{vec(\mathbf{P}_1)^t vec(\mathbf{P}_1)}\sqrt{vec(\mathbf{P}_2)^t vec(\mathbf{P}_2)}}$$

is a correlation between vectorized $\mathbf{P}_1$ and $\mathbf{P}_2$ introduced in Escoufier (1973) as the RV coefficient.

Proposition 1 implies that, for nonzero k_1 and k_2, the distances $D^2_w(\mathbf{P}_1, \mathbf{P}_2)$ get any values on the closed intervals

$$(a) : \left[\frac{1}{2}|k_1 - k_2|, \frac{1}{2}(k_1 + k_2) + \min\{p - k_1 - k_2, 0\}\right],$$

$$(b) : \left[\frac{1}{2}\left|k_1^{-1} - k_2^{-1}\right|, \frac{1}{2}\left(k_1^{-1} + k_2^{-1}\right) + k_1^{-1}k_2^{-1}\min\{p - k_1 - k_2, 0\}\right], \text{ and}$$

$$(c) : \left[1 - \min\{k_1^{1/2}k_2^{-1/2}, k_1^{-1/2}k_2^{1/2}\}, 1 + k_1^{-1/2}k_2^{-1/2}\min\{p - k_1 - k_2, 0\}\right].$$

If $k_1 = 0$, for example, then $D^2_w(\mathbf{P}_1, \mathbf{P}_2)$ is simply $w^2(k_2)k_2/2$. Recall that, for all three choices of weights, the distance is zero only if $\mathbf{P}_1 = \mathbf{P}_2$ (and $k_1 = k_2$). For weights w_a, the largest possible value for $D^2_w(\mathbf{P}_1, \mathbf{P}_2)$ is $p/2$ and it is obtained if and only if $\mathbf{P}_1$ and $\mathbf{P}_2$ are orthogonal and $\mathbf{P}_1 + \mathbf{P}_2 = \mathbf{I}_p$ (i.e., $k_1 + k_2 = p$). For weights w_b, $D^2_w(\mathbf{P}_1, \mathbf{P}_2) \leq 1$, and $D^2_w(\mathbf{P}_1, \mathbf{P}_2) = 1$ if and only if $\mathbf{P}_1$ and $\mathbf{P}_2$ are orthogonal and $k_1 = k_2 = 1$. Finally, for weights w_c, the maximum value $D_w(\mathbf{P}_1, \mathbf{P}_2) = 1$ for $k_1, k_2 \neq 0$ is attained as soon as $\mathbf{P}_1$ and $\mathbf{P}_2$ are orthogonal and $k_1 + k_2 \leq p$.

The following two special cases illustrate the differences between the three distances.

1. First, consider the case when $\mathscr{S}_{\mathbf{P}_1} \subset \mathscr{S}_{\mathbf{P}_2}$. Then naturally $tr(\mathbf{P}_1\mathbf{P}_2) = tr(\mathbf{P}_1) = k_1$ and

$$D^2_w(\mathbf{P}_1, \mathbf{P}_2) = \frac{w^2(k_1)k_1 + w^2(k_2)k_2}{2} - w(k_1)w(k_2)k_1$$

and therefore, for $k_2 \neq 0$ and with $\lambda = k_1/k_2$

$$D^2_{w_a}(\mathbf{P}_1, \mathbf{P}_2) = \frac{k_2}{2}(1 - \lambda),$$

$$D^2_{w_b}(\mathbf{P}_1, \mathbf{P}_2) = \frac{1}{2k_1}(1 - \lambda), \text{ and}$$

$$D^2_{w_c}(\mathbf{P}_1, \mathbf{P}_2) = 1 - \sqrt{\lambda}.$$

One can see that $D^2_{w_c}(\mathbf{P}_1, \mathbf{P}_2)$ depends only on the ratio between k_1 and k_2, which can be seen as a nice feature. $D^2_{w_a}(\mathbf{P}_1, \mathbf{P}_2)$ and $D^2_{w_b}(\mathbf{P}_1, \mathbf{P}_2)$ however depend additionally on the actual values of k_1 and k_2. In fact, $D^2_{w_a}(\mathbf{P}_1, \mathbf{P}_2)$ depends on the difference $k_2 - k_1$. It implies that the distance between the subspace $\mathscr{S}_{\mathbf{P}_1}$ and $\mathscr{S}_{\mathbf{P}_2}$ will remain the same whether the dimensions $k_1 = 1$ and $k_2 = 2$ or whether $k_1 = 99$ and $k_2 = 100$, while we would expect the distance to be much smaller in the latter case.

2. Second, consider the case when $\mathscr{S}_{\mathbf{P}_1}$ and $\mathscr{S}_{\mathbf{P}_2}$ are orthogonal, that is, when $tr(\mathbf{P}_1\mathbf{P}_2) = 0$. Then

$$D_w^2(\mathbf{P}_1, \mathbf{P}_2) = \frac{w^2(k_1)k_1 + w^2(k_2)k_2}{2}$$

and therefore, for nonzero k_1 and k_2

$$\begin{aligned} D_{w_a}^2(\mathbf{P}_1, \mathbf{P}_2) &= \frac{1}{2}(k_1 + k_2), \\ D_{w_b}^2(\mathbf{P}_1, \mathbf{P}_2) &= \frac{1}{2}\left(\frac{1}{k_1} + \frac{1}{k_2}\right), \quad \text{and} \\ D_{w_c}^2(\mathbf{P}_1, \mathbf{P}_2) &= 1. \end{aligned}$$

It is natural to think subspaces that are orthogonal to each other are furthest apart possible. This information is apparent in $D_{w_c}^2(\mathbf{P}_1, \mathbf{P}_2)$. However, interpreting both $D_{w_a}^2(\mathbf{P}_1, \mathbf{P}_2)$ and $D_{w_b}^2(\mathbf{P}_1, \mathbf{P}_2)$ is again more difficult since they depend on the actual values of k_1 and k_2.

2.3 Averages of Subspaces with Arbitrary Dimensions

Consider the orthogonal projections $\mathbf{P}_1, \ldots, \mathbf{P}_m$ with ranks $k_1, \ldots, k_m$. To combine the orthogonal projections we give the following

Definition 2 The average orthogonal projection (AOP) $\mathbf{P}_w$ based on the weights $w(0), \ldots, w(p)$ is an orthogonal projection that minimizes the objective function

$$\sigma_w^2(\mathbf{P}) = \frac{1}{m}\sum_{i=1}^{m} D_w^2(\mathbf{P}_i, \mathbf{P}).$$

To find the AOP, we can use the following result.

Lemma 2 *The AOP $\mathbf{P}_w$ maximizes the function*

$$D(\mathbf{P}) = w(k)tr(\bar{\mathbf{P}}_w\mathbf{P}) - \frac{1}{2}w^2(k)k,$$

where

$$\bar{\mathbf{P}}_w = \frac{1}{m}\sum_{i=1}^{m} w(k_i)\mathbf{P}_i$$

is a regular average of weighted orthogonal projections, and k is the rank of **P**.

Proof As shown before

$$D_w^2(\mathbf{P}_i, \mathbf{P}) = \frac{1}{2}w^2(k_i)k_i + \frac{1}{2}w^2(k)k - w(k_i)w(k)tr(\mathbf{P}_i\mathbf{P}).$$

Then

$$\sigma_w^2(\mathbf{P}) = \frac{1}{m}\sum_{i=1}^{m} D_w^2(\mathbf{P}_i, \mathbf{P}) = \frac{1}{2m}\sum_{i=1}^{m} w^2(k_i)k_i + \frac{1}{2}w^2(k)k - w(k)tr(\bar{\mathbf{P}}_w\mathbf{P}).$$

The first term in the latest sum does not depend on **P** or k. Thus, $\sigma_w^2(\mathbf{P})$ is minimized when $w(k)tr(\bar{\mathbf{P}}_w\mathbf{P}) - \frac{1}{2}w^2(k)k$ is maximized. □

Naturally, $\bar{\mathbf{P}}_w$ is symmetric and nonnegative definite, but not an orthogonal projection anymore. In the following derivations, we need its eigenvector and eigenvalue decomposition

$$\bar{\mathbf{P}}_w = \mathbf{U}\boldsymbol{\Lambda}\mathbf{U}^t = \sum_{i=1}^{p} \lambda_i \mathbf{u}_i\mathbf{u}_i^t,$$

where $\lambda_1 \geq \ldots \geq \lambda_p \geq 0$ and $\mathbf{u}_i$ is the eigenvector corresponding to the eigenvalue λ_i. Recall that the eigenvectors are uniquely defined only for eigenvalues that are distinct from other eigenvalues. Using the Lemma 2 and the eigenvector and eigenvalue decomposition $\bar{\mathbf{P}}_w$, our main result easily follows.

Proposition 2 *The rank k of the AOP* $\mathbf{P}_w$ *maximizes the function*

$$f_w(k) = w(k)(\sum_{i=1}^{k} \lambda_i)I(k > 0) - \frac{1}{2}w^2(k)k, \quad k = 0, \ldots, p,$$

where $\lambda_1 \geq \ldots \geq \lambda_p \geq 0$ *are the eigenvalues of* $\bar{\mathbf{P}}_w$. *Moreover*

$$\mathbf{P}_w = I(k > 0) \cdot \sum_{i=1}^{k} \mathbf{u}_i\mathbf{u}_i^t$$

where $u_1, \ldots, u_k$ *are the eigenvectors corresponding to eigenvalues* $\lambda_1, \ldots, \lambda_k$.

Proof The AOP $\mathbf{P}_w$ maximizes

$$D(\mathbf{P}) = w(k)tr(\bar{\mathbf{P}}_w\mathbf{P}) - \frac{1}{2}w^2(k)k,$$

where k is the rank of $\mathbf{P}$. Assume first that $k > 0$ is fixed and $\mathbf{P} = \mathbf{V}\mathbf{V}^t$ where $\mathbf{V}$ has k orthonormal columns. Then $D(\mathbf{P})$ is maximized as soon as $tr(\bar{\mathbf{P}}_w\mathbf{P})$ is maximized. Then, as $\bar{\mathbf{P}}_w = \sum_{i=1}^p \lambda_i \mathbf{u}_i\mathbf{u}_i^t$, $tr(\bar{\mathbf{P}}_w\mathbf{P}) = tr(\bar{\mathbf{P}}_w\mathbf{V}\mathbf{V}^t) = tr(\mathbf{V}^t\bar{\mathbf{P}}_w\mathbf{V})$ is maximized if $\mathbf{V} = (\mathbf{u}_1, ..., \mathbf{u}_k)$, and the maximum value is $\sum_{i=1}^k \lambda_i$. For fixed $k > 0$, the maximum value of $D(\mathbf{P})$ is then $w(k)\sum_{i=1}^k \lambda_i - \frac{1}{2}w^2(k)k$, and $D(0) = 0$. The result follows. □

Note that the calculation of the AOP $\mathbf{P}_w$ is easy, only the eigenvalues and eigenvectors of $\bar{\mathbf{P}}_w$ are needed. The AOP $\mathbf{P}_w$ is not always unique. This happens for example if the rank of an AOP is k and $\lambda_{k+1} = \lambda_k$. Consider now the three weight functions

$$(a)\ w_a(k) = 1, \quad (b)\ w_b(k) = \frac{1}{k}, \quad \text{and } (c)\ w_c(k) = \frac{1}{\sqrt{k}}, \quad \text{for } k > 0.$$

The function f_w for these three weight functions is, for $k > 0$

$$(a) \sum_{i=1}^k \lambda_i - \frac{k}{2}, \quad (b)\ \frac{1}{k}\left(\sum_{i=1}^k \lambda_i - \frac{1}{2}\right), \quad \text{and } (c)\ \frac{1}{\sqrt{k}}\left(\sum_{i=1}^k \lambda_i\right) - \frac{1}{2}.$$

Note that $f_w(0) = 0$ for all weights w. To find local maxima for these functions, one can then use the results

$$(a): f_w(k+1) \geq f_w(k) \Leftrightarrow \lambda_{k+1} \geq \frac{1}{2},$$

$$(b): f_w(k+1) \geq f_w(k) \Leftrightarrow \lambda_{k+1} \geq \frac{1}{k}\left(\lambda_1 + \ldots + \lambda_k - \frac{1}{2}\right), \quad \text{and}$$

$$(c): f_w(k+1) \geq f_w(k) \Leftrightarrow \lambda_{k+1} \geq \left(\sqrt{\frac{k+1}{k}} - 1\right)(\lambda_1 + \ldots + \lambda_k)$$

for $k = 1, \ldots, p-1$.

Note that for (a), $f_w(k)$ is a concave function and the global maximum is simply the largest k with the eigenvalue $\lambda_k \geq \frac{1}{2}$. The functions in (b) and (c) are not concave, however, and the global maximum is found by computing all the values $f_w(k)$, $k = 0, ..., p$.

3 Application

In this section we discuss the performance of the average orthogonal projections (AOP) for four different dimension reduction problems. The orthogonal projections and their combinations aim for different targets in different applications. Each problem with natural orthogonal projections is first shortly introduced, and then the performance of AOP is demonstrated using three simulation studies and a real data example. The three simulated studies lead to different situations concerning the estimation of ranks. In the first simulated example, the ranks of the orthogonal projections together with the rank of the AOP are considered as fixed while only the rank of the AOP is estimated in the second example and all ranks are estimated in the third example. The computations in this section are done using R (R Development Core Team 2012) and the packages bootstrap (Tibshirani 2013), class (Venables and Ripley 2002), dr (Weisberg 2002), ICS (Nordhausen et al. 2008), MASS (Venables and Ripley 2002), MMST (Halbert 2011), MNM (Nordhausen and Oja 2011), pcaPP (Filzmoser et al. 2012) and robustbase (Rousseeuw et al. 2012). Furthermore, the weighted distance (Definition 1) and the AOP (Definition 2) are implemented in the R package LDRTools (Liski et al. 2014b).

3.1 Principal Component Analysis

Classical principal component analysis (PCA) may be based on the eigenvector and eigenvalue decomposition of the covariance matrix of a p-variate random vector $\mathbf{x}$, that is, on

$$cov(\mathbf{x}) = \mathbf{U}\Lambda\mathbf{U}^t = \sum_{i=1}^{p} \lambda_i \mathbf{u}_i \mathbf{u}_i^t$$

where $\lambda_1 \geq ...\lambda_p \geq 0$ are the ordered eigenvalues and $\mathbf{u}_1, ..., \mathbf{u}_p$ are the corresponding eigenvectors. The orthogonal projection $\mathbf{P}_{cov} = \sum_{i=1}^{k} \mathbf{u}_i \mathbf{u}_i^t$ then projects p-variate observations to the k-variate subset with maximum variation. It is unique if $\lambda_{k+1} > \lambda_k$.

Let $F_{\mathbf{x}}$ be the cumulative distribution function of $\mathbf{x}$. A $p \times p$ matrix valued functional $S(F_{\mathbf{x}})$ is a scatter matrix if $S(F_{\mathbf{x}})$ is a nonnegative definite and symmetric matrix with the affine equivariance property

$$S(F_{\mathbf{Ax+b}}) = \mathbf{A}S(F_{\mathbf{x}})\mathbf{A}^t \text{ for all full-rank } p \times p \text{ matrices } \mathbf{A} \text{ and all } p\text{-vectors } \mathbf{b}.$$

It is remarkable that, if $\mathbf{x}$ has an elliptic distribution then the ordered eigenvectors of $S(F_{\mathbf{x}})$ are those of $cov(\mathbf{x})$. Therefore, in the elliptic case, any scatter matrix can be used to find $\mathbf{P} = \sum_{i=1}^{k} \mathbf{u}_i \mathbf{u}_i^t$ and the matrix $\mathbf{P}$ is a well-defined population quantity even if the second moments (and the covariance matrix) do not exist. Naturally, the sample statistics corresponding to different scatter matrices then have different statistical (efficiency and robustness) properties. For a fixed value of k, one can then try to "average" these different PCA approaches to get a compromise estimate.

We next illustrate the performance of the AOP in the following simple scenario. Let first $\mathbf{x} \sim N_6(0, \boldsymbol{\Lambda})$, $\boldsymbol{\Lambda} = diag(9, 7, 5, 1, 1, 1)$. We choose $k = 3$ and wish to estimate $\mathbf{P}_{\boldsymbol{\Lambda}} = diag(1, 1, 1, 0, 0, 0)$. Let then $x_1, ..., x_n$ be a random sample from $N_6(0, \boldsymbol{\Lambda})$, and find an estimate $\mathbf{P}_{cov}$, an orthogonal projection with rank $k = 3$ obtained from the sample covariance matrix. This estimate is then combined with three robust estimates, namely,

- $\mathbf{P}_{Tyler}$ that is based on Tyler's shape matrix (Tyler 1987) with the affine equivariant version of spatial median as a multivariate location estimate (Hettmansperger and Randles 2002).
- $\mathbf{P}_{MCD}$ that is based on the minimum covariance determinant (MCD) estimator (Rousseeuw 1986).
- $\mathbf{P}_{PP}$ that is based on the projection pursuit (PP) approach for PCA with the median absolute deviation (mad) criterion as suggested in Croux and Ruiz-Gazen (2005).

In the simulations, $\mathbf{x}_1, \ldots, \mathbf{x}_n$ was a random sample from $N_6(0, \boldsymbol{\Lambda})$ with $n = 400$, and the sampling was repeated 1000 times. As $k_1 = \ldots = k_m = k = 3$ is fixed, the AOP $\mathbf{P}_w$ does not depend on the weight function and we use only w_a. A similar simulation study was conducted but with observations coming from a heavy-tailed elliptical t_2 distribution with the corresponding scatter matrix $\boldsymbol{\Lambda} = diag(9, 7, 5, 1, 1, 1)$. Note that the regular covariance matrix does not exist in this case but the true orthogonal projection is still well defined. The average squared distances $D^2_{w_a}$ between the four orthogonal projection estimates, their AOP $\mathbf{P}_{w_a}$, and $\mathbf{P}_{\boldsymbol{\Lambda}}$ are shown in Table 1 for both the normal and the t-distributed settings, respectively in the left and right columns. Note that for both settings $\mathbf{P}_{\boldsymbol{\Lambda}}$ is the same, but the four orthogonal projection estimates and their AOP differ.

The results in the multivariate normal case show, as expected, that the orthogonal projection estimate based on the covariance matrix is the best one here. Also the average orthogonal projection performs very well although it combines information coming from the much worse $\mathbf{P}_{PP}$. In the t_2 distribution case, traditional $\mathbf{P}_{cov}$ fails but the average orthogonal projection is still performing well. Recall the second moments and $\mathbf{P}_{cov}$ do not exist in this case.

Table 1 Average squared distances $D^2_{w_a}$ between the four orthogonal projection estimates, their AOP $\mathbf{P}_{w_a}$, and true $\mathbf{P}_{\boldsymbol{\Lambda}}$. For all orthogonal projections, rank $k = 3$. The averages are based on 1000 random samples of size $n = 400$ from $N_6(0, \boldsymbol{\Lambda})$ (*left column*) and t_2 (*right column*)

	$\mathbf{P}_{\boldsymbol{\Lambda}}(N)$	$\mathbf{P}_{\boldsymbol{\Lambda}}(t)$
$\mathbf{P}_{cov}$	0.005	0.114
$\mathbf{P}_{Tyler}$	0.007	0.007
$\mathbf{P}_{MCD}$	0.007	0.012
$\mathbf{P}_{PP}$	0.061	0.070
$\mathbf{P}_{w_a}$	0.009	0.016

In practice, the distribution of the random sample is unknown, but from the simulations we can see that whether the distribution is gaussian or with heavy tails, the AOP procedure will make the most of the different PCA methods.

3.2 Averaging One-Dimensional PP Orthogonal Projections

In the previous section we used the projection pursuit (PP) approach for principal component analysis. PP is a much more general technique, however, and there are many other types of indices than just measures of variation to define "interesting" one-dimensional directions. PP actually dates back to Friedman and Tukey (1974) and usually one searches for non-gaussian directions. For a recent review of the existing indices, see for example Rodriguez-Martinez et al. (2010). A major challenge in PP is that it is computationally difficult to find the direction which globally maximizes the index and that there are usually several local maxima. However, since the local maxima may be also of interest, one possible strategy, as detailed in Ruiz-Gazen et al. (2010), is to run the algorithm many times using different initializations. With this strategy, the user has many orthogonal projections of rank one but many of them are usually redundant. So, it is of particular interest to summarize all these orthogonal projections in order to extract the directions that are useful and unique. It means that, in that case, one is interested in an average orthogonal projection of orthogonal projections with rank one that may have a higher rank.

To demonstrate the interest of AOP in the context of PP, we choose the deflation-based fastICA method (Hyvärinen 1999) as an example since it is well understood and computationally quite simple. While deflation-based fastICA is originally developed in the context of independent component analysis (ICA), it can be seen as a traditional PP approach when only one direction is extracted. For a random variable $\mathbf{x}$ with the standardized version $\mathbf{z} = cov(\mathbf{x})^{-1/2}(\mathbf{x} - E(\mathbf{x}))$, deflation-based fastICA maximizes a measure of non-gaussianity of the form $|E(G(\mathbf{u}^t\mathbf{z}))|$, under the constraint that $\mathbf{u}^t\mathbf{u} = 1$, where G is a selected twice differentiable nonlinear nonquadratic function with $G(0) = 0$. The final PP direction is then $(\mathbf{u}^t cov(\mathbf{x})^{-1}\mathbf{u})^{-1/2}\, cov(\mathbf{x})^{-1/2}\mathbf{u}$, and the corresponding orthogonal projection is $(\mathbf{u}^t cov(\mathbf{x})^{-1}\mathbf{u})^{-1}\ cov(\mathbf{x})^{-1/2}\,\mathbf{u}\mathbf{u}^t cov(\mathbf{x})^{-1/2}$. In our simulations, we use four common choices of $G(\mathbf{u})$ with derivative functions $g(\mathbf{u})$: (i) $\mathbf{u}^3$, (ii) $\tanh(\mathbf{u})$, (iii) $\mathbf{u}\exp(\mathbf{u}^2/2)$, and (iv) $\mathbf{u}^2$. If there are more than one non-gaussian direction in the data, the direction to be found depends heavily on the initial value of the algorithm, see e.g. Nordhausen et al. (2011), Miettinen et al. (2014).

In our simulation study, we choose a 10-variate $\mathbf{x} = (x_1, \ldots, x_{10})^t$ where the first three variables are mixtures of gaussian distributions and $x_i \sim N(0, 1)$, for $i = 4, \ldots, 10$. More precisely, $x_1 = \frac{1}{\sqrt{5}}(p_1 y_1 + (1 - p_1)y_2 - 2)$ with $p_1 \sim Bin(1, 0.5)$, $y_1 \sim N(0, 1)$ and $y_2 \sim N(4, 1)$; $x_2 = 1/\sqrt{2.89}(p_2 y_3 + (1 - p_2)y_4 - 2.1)$ with $p_2 \sim Bin(1, 0.3)$, $y_3 \sim N(0, 1)$ and $y_4 \sim N(3, 1)$, and $x_3 = 1/\sqrt{24.36}(p_3 y_5 + (1 - p_3)y_6 - 2)$ with $p_3 \sim Bin(1, 0.4)$, $y_5 \sim N(0, 9)$ and $y_6 \sim N(8, 9)$. We generated 1000 ran-

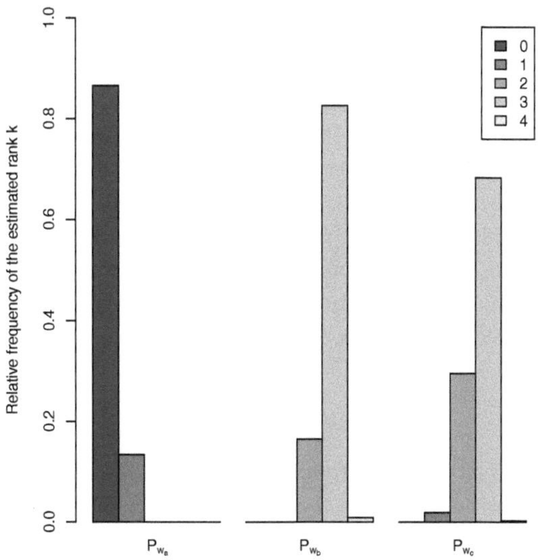

Fig. 1 Relative frequencies of the estimated ranks of the AOP using the weight functions w_a, w_b and w_c

dom samples of sizes $n = 200$ from the 10-variate distribution described above. For each sample, we found 100 one-dimensional PP directions (4 choices of G, 25 random initial values for the algorithm for each choice of G). For each sample, 100 PP orthogonal projections were then averaged using each of the three weight functions w_a, w_b, and w_c. In this setting, the average orthogonal projection should be close to the orthogonal projection $\mathbf{P}_{true} = diag(1, 1, 1, 0, \ldots, 0)$ with rank 3 that picks the three non-gaussian components of the data.

Figure 1 shows the relative frequencies of the ranks of the AOPs obtained with the three weight functions in 1000 repetitions. Clearly the weight function w_a is not appropriate in this application because $k_1 = k_2 = \ldots = k_m = 1$ implies that $\sum_{i=1}^{m} \lambda_i = 1$ with $\lambda_i \geq 0$, which means that there cannot be more than one eigenvalue larger than 1/2 and, consequently, the rank k equals zero or one. With weight functions w_b and w_c, the correct rank 3 is obtained in 82.6 % and 68.3 % of the runs, respectively. It is also hoped that the AOPs are close to the true orthogonal projection $\mathbf{P}_{true}$. To evaluate that, we calculated in Table 2 the average distances $D^2_{w_a}$, $D^2_{w_b}$, and $D^2_{w_c}$ (columnwise) between $\mathbf{P}_{w_a}$ and $\mathbf{P}_{true}$, between $\mathbf{P}_{w_b}$ and $\mathbf{P}_{true}$, and between $\mathbf{P}_{w_c}$ and $\mathbf{P}_{true}$. Notice that, for all distances, the AOP $\mathbf{P}_{w_b}$ is closest on average to the true value $\mathbf{P}_{true}$, which illustrates the interest of averaging.

Table 2 Average squared distances $D^2_{w_a}$, $D^2_{w_b}$, and $D^2_{w_c}$ (columnwise) between $\mathbf{P}_{true}$ and $\mathbf{P}_{w_a}$, $\mathbf{P}_{w_b}$ and $\mathbf{P}_{w_c}$, respectively

	$\mathbf{P}_{true}(D^2_{w_a})$	$\mathbf{P}_{true}(D^2_{w_b})$	$\mathbf{P}_{true}(D^2_{w_c})$
$\mathbf{P}_{w_a}$	1.005	0.335	0.425
$\mathbf{P}_{w_b}$	0.122	0.018	0.043
$\mathbf{P}_{w_c}$	0.199	0.035	0.074

3.3 Supervised Dimension Reduction

In the PCA application, we used the same $k = 3$ for the orthogonal projections and their AOP. In the PP application, the rank of the orthogonal projections was taken as one while the rank of their AOP was not fixed. However, for many dimension reduction methods, the ranks of the individual orthogonal projections are not fixed but also estimated from the data, and the ranks may differ from one method to another. We now look at this scenario in the framework of supervised dimension reduction.

In supervised dimension reduction, one often assumes that a response variable y and the p-vector $\mathbf{x}$ are related through

$$y = f(\mathbf{b}_1^t\mathbf{x}, \ldots, \mathbf{b}_k^t\mathbf{x}, \varepsilon),$$

with an unknown function f and an unknown error variable ε. The goal of supervised dimension reduction is to estimate the value of k and the matrix $\mathbf{B} = (\mathbf{b}_1, \ldots, \mathbf{b}_k)$ to obtain $\mathbf{P_B}$ with rank k. Hence, for supervised dimension reduction, the joint distribution of y and $\mathbf{x}$ is of interest and, for the matrix $\mathbf{B}$, it holds that $y \perp\!\!\!\perp \mathbf{x}|\mathbf{B}^t\mathbf{x}$.

Many supervised dimension reduction methods have been suggested in the literature and their performances often strongly depend on the unknown function f. The well-known sliced inverse regression (SIR) for example may not find directions with nonlinear dependencies while, on the other hand, principal Hessian directions (PHD) cannot find linear relationships. Hence, when using supervised dimension reduction methods in practice, the estimated rank k and the corresponding orthogonal projection might differ considerably depending on the method. Therefore there were already several approaches suggested in the framework of supervised dimension reduction to combine different methods, see for example Ye and Weiss (2003), Shaker and Prendergast (2011) and references therein for further details. We propose to use in the following AOP in order to summarize in an efficient way the information brought by the complementary estimation strategies.

In our example we generate data sets from the following two models:

$$\begin{aligned} \text{M1:} \quad y &= \mathbf{b}_{11}^t\mathbf{x} - (\mathbf{b}_{12}^t\mathbf{x})^2 + \sigma\varepsilon \\ \text{M2:} \quad y &= (\mathbf{b}_{21}^t\mathbf{x})^2 + \sigma\varepsilon, \end{aligned}$$

where $\mathbf{x} \sim N_{10}(0, \mathbf{I}_{10})$, $\varepsilon \sim N(0, 1)$, $\sigma = 0.5$ and $\mathbf{b}_{ij}$'s are all 10-dimensional column vectors

$$\begin{aligned} \mathbf{b}_{11} &= (2, 3, 0, \ldots, 0)^t \text{ and } \mathbf{b}_{12} = (0, 0, \sqrt{5}, 0, \ldots, 0)^t, \\ \mathbf{b}_{21} &= (1, 0, 0, 0, \ldots, 0)^t. \end{aligned}$$

Hence $k = 2$ for model M1 and $k = 1$ for model M2. In both cases, we generated 100 samples of size 400.

In our illustration, we use supervised dimension reduction methods implemented in the dr package that provide both the estimate of k and the corresponding orthogonal projection estimate. The estimation strategies are then (i) sliced inverse regression (SIR), (ii) sliced variance estimation (SAVE), (iii) inverse regression estimation (IRE), and three types of principal hessian directions (PHD), namely, (iv) response based principal hessian directions (PHDY), (v) residual based principal hessian directions (PHDR), and (vi) the so-called q-based principal hessian directions (PHDQ). For details about these estimation methods, see Weisberg (2002) and references therein. We also add here (vii) PCA with k chosen simply as the number of eigenvalues larger than 1. Naturally, PCA ignores y and is therefore not supervised. (Its use could be motivated by the aim to avoid directions with small variation. In our case it just provides random orthogonal projections.) Furthermore, do we compare our proposal with another combination approach (viii) suggested by Ye and Weiss (2003) called bootstrapping (BOOT) that also combines several reduction dimension methods. Roughly speaking, the method is based on linear combinations of matrices obtained from different dimension reduction methods that lead to estimators of the subspace and its dimension. The methods we combine in this study are the so-called SIR and SAVE matrices M_{SIR} and M_{SAVE} using $(1-c)M_{SIR} + cM_{SAVE}$, $c = 0, 0.1, \ldots, 1$. Among all the possible linear combinations we find the linear combination of minimal variability using the vector correlation coefficient as measure. Ye and Weiss (2003) presented also an approach for estimating k, but due to the tedious task of the process we fix k to the true in our simulation study. For more details on the bootstrapping combination approach, see Ye and Weiss (2003).

In the following we compare the eight methods above and the AOPs for methods (i)–(vii) based on the weight functions w_b and w_c. The use of w_a is not reasonable here because of the varying estimates of k. We consider therefore here the following two AOPs.

AOP1: The AOP using w_b with estimated k.
AOP2: The AOP using w_c with estimated k.

Some simulation results are collected in Figs. 2, 3. The figures show the boxplots for the observed $D^2_{w_b}$ and $D^2_{w_c}$ distances between the true orthogonal projection and the orthogonal projection estimates coming from the different dimension reduction

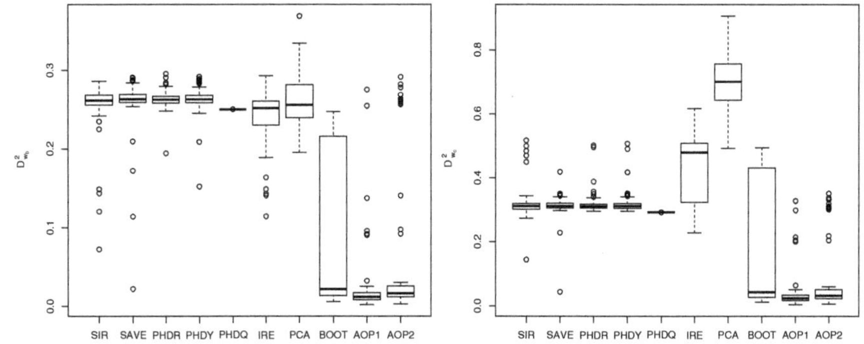

Fig. 2 Boxplots of the observed $D^2_{w_b}$ (*left panel*) and $D^2_{w_c}$ (*right panel*) distances between the true and estimated orthogonal projections when the observations come from the model M1 with $k = 2$

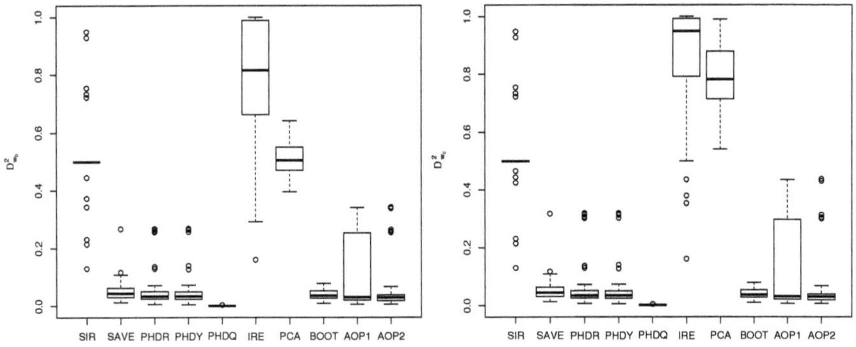

Fig. 3 Boxplots of the observed $D^2_{w_b}$ (*left panel*)and $D^2_{w_c}$ (*right panel*) distances between the true and estimated orthogonal projections when the observations come from the model M2 with $k = 1$

approaches. Consider first the behavior of the estimates in the model M1 with $k = 2$ (see Fig. 2). The performances of SIR, SAVE and PHD estimates seem to be very similar and they usually find only one direction. (For example, SIR finds only the component with linear dependence, and SAVE only the component of quadratic dependence.) The same seems to be true with IRE but with more varying estimates. Recall that, if $\mathscr{S}_{\mathbf{P}_1} \subset \mathscr{S}_{\mathbf{P}_2}$ and $k_1 = 1$ and $k_2 = 2$, then $\lambda = k_1/k_2 = 0.5$, and the average distances of SIR, SAVE and PHD estimates tend to be close to

$$D^2_{w_b}(\mathbf{P}_1, \mathbf{P}_2) = \frac{1}{2k_1}(1 - \lambda) = 0.25 \;\text{ and }\; D^2_{w_c}(\mathbf{P}_1, \mathbf{P}_2) = 1 - \sqrt{\lambda} = 0.293,$$

respectively. The AOP estimates nicely pick up the two dimensions and clearly outperform other estimates. Note that even though k is fixed and true for BOOT, it is still outperformed by the AOP estimates. PCA has a poor performance as expected.

Figure 3 gives the results for the model M2 with $k = 1$ and quadratic dependence. SAVE and PHD approaches work very well, and SIR and IRE completely fail in this case. AOP using w_c neglects the bad estimates and picks up nicely the correct direction, whereas AOP using w_b is somewhat affected by the bad estimates. BOOT performs very well and as in model M1, PCA provides a random reference method with a bad performance.

4 Real Data Example

As the final example AOP is demonstrated on a real data set which considers pen-based recognition of handwritten digits. The data set which is available in the R package MMST contains 16 variables for 10992 digits ranging from 0–9 and the usual goal is to perform supervised classification. In our illustration we first divide the data set randomly into two equal parts, training data and testing data, we next use different dimension reductions methods and two different supervised classification methods with training data to obtain classification rules, and the rules are then applied to the testing data where the misclassification percentage is computed.

As dimension reduction methods we use all four principal component analysis methods as described in Sect. 3.1 denoted in the following as PCA_{cov}, PCA_{Tyler}, PCA_{MCD} and PCA_{PP} respectively. For each method the dimension is chosen visually using screeplots. Furthermore we use two versions of invariant coordinate selection (ICS) (Tyler et al. 2009). The first version denoted as ICS_{cov,cov_4} uses the regular covariance matrix and the matrix of fourth moments whereas the second version denoted $ICS_{MCD,Tyler}$ uses MCD and Tyler's shape matrix. In both cases the dimensions are chosen visually based on scatter plots. The last dimension reduction method is SIR, which can be seen as a supervised version of ICS, see Liski et al. (2014a). In SIR we use 10 slices corresponding to the 10 digits and the estimated dimension is based on the tests available in the dr package. All these methods are then combined using the three weight functions introduced earlier with their own estimate for the dimension. Note that the dimension reduction methods here are quite different. The principal components methods search for directions with large variation mainly having an elliptical model in mind whereas the ICS methods look for rather non-elliptical directions with extreme kurtosis values.

For all 10 dimension reduction methods classification rules are then built using 1-nearest neighbor (knn) classification and quadratic discriminant analysis (QDA) and evaluated on the testing data. Table 3 gives for all 10 different dimension reduction methods and the two different classifiers the estimated dimension of the data and the percentage of misclassification in the testing data. The table shows that for the individual dimension reduction methods the estimated dimension of the data varies from 3 to 5, where PCA_{PP} with an estimate of 4 yields the smallest classification error for both knn and QDA. When combining these individual methods the AOP with weight function w_b prefers a rank higher than any of the individual methods and

Table 3 Classification errors in percent and the estimated dimensions k for the different dimension reduction and classification methods

MET	k	knn	QDA
PCA_{cov}	3	0.14	0.18
PCA_{Tyler}	3	0.16	0.20
PCA_{MCD}	3	0.20	0.25
PCA_{PP}	4	0.08	0.13
ICS_{cov,cov_4}	5	0.21	0.28
$ICS_{MCD,Tyler}$	5	0.20	0.22
SIR	4	0.13	0.19
AOPw_a	3	0.17	0.22
AOPw_b	6	0.04	0.04
AOPw_c	5	0.07	0.12

gives in total the smallest classification error. But also the weight function w_c which uses as rank 5 still obtains classification errors smaller than any of the individual methods which makes very clear that for these data different methods emphasize different features of the data and the average of all of them is most informative for further analysis.

5 Final Comments

Dimension reduction and subspace estimation is a topic with increasing relevance since modern data sets become larger and larger. Different approaches have different shortcomings and combining the results coming from different approaches might give a better total overview. In this paper, we propose a generalization of the Crone and Crosby distance for the orthogonal projections, a weighted distance that allows to combine subspaces of different dimensions. Some natural choices of weights are considered in detail. The performance of three weighted distances and the combining approach is illustrated via simulations and a real data example, which show that each of them has its own justification depending on the problem at hand. Similar to other areas of statistics, this kind of "model averaging" seems to be a way to combine information from competing estimates and to give a better idea of the true model at hand.

Acknowledgments The work of Klaus Nordhausen and Hannu Oja was supported by the Academy of Finland (grant 268703). The authors are grateful to the reviewers for their helpful comments.

References

Cook RD, Weisberg S (1991) Sliced inverse regression for dimension reduction: comment. J Am Stat Assoc 86:328–332

Crone LJ, Crosby DS (1995) Statistical applications of a metric on subspaces to satellite meteorology. Technometrics 37:324–328

Croux C, Ruiz-Gazen A (2005) High breakdown estimators for principal components: the projection-pursuit approach revisited. J Multivar Anal 95:206–226

Escoufier Y (1973) Le traitement des variables vectorielles. Biometrics 29:751–760

Filzmoser P, Fritz H, Kalcher K (2012) pcaPP: Robust PCA by projection pursuit. R package version 1.9-47

Friedman JH, Tukey JW (1974) A projection pursuit algorithm for exploratory data analysis. IEEE Trans Comput C 23:881–889

Halbert K (2011) MMST: Datasets from MMST. R package version 0.6-1.1

Hettmansperger TP, Randles RH (2002) A practical affine equivariant multivariate median. Biometrika 89:851–860

Hotelling H (1936) Relations between two sets of variates. Biometrika 28:321–377

Hyvärinen A (1999) Fast and robust fixed-point algorithms for independent component analysis. IEEE Trans Neural Netw 10:626–634

Li KC (1991) Sliced inverse regression for dimension reduction. J Am Stat Assoc 86:316–327

Li KC (1992) On principal Hessian directions for data visualization and dimension reduction: another application of Stein's lemma. J Am Stat Assoc 87:1025–1039
Liski E, Nordhausen K, Oja H (2014a) Supervised invariant coordinate selection. Stat: A J Theoret Appl Stat 48:711–731
Liski E, Nordhausen K, Oja H, Ruiz-Gazen A (2014b) LDRTools: tools for linear dimension reduction. R package version 1
Miettinen J, Nordhausen K, Oja H, Taskinen S (2014) Deflation-based FastICA with adaptive choices of nonlinearities. IEEE Trans Signal Process 62:5716–5724
Nordhausen K, Oja H, Tyler DE (2008) Tools for exploring multivariate data: the package ICS. J Stat Soft 28(6):1–31
Nordhausen K, Ilmonen P, Mandal A, Oja H, Ollila E (2011) Deflation-based FastICA reloaded. Proceedings of 19th European signal processing conference 2011 (EUSIPCO 2011) 1854–1858
Nordhausen K, Oja H (2011) Multivariate L1 methods: the package MNM. J Stat Softw 43:1–28
Development Core Team R (2012) R: A language and environment for statistical computing. R Foundation for Statistical Computing, Vienna, Austria
Rodriguez-Martinez E, Goulermas JY, Mu T, Ralph JF (2010) Automatic induction of projection pursuit indices. IEEE Trans Neural Netw 21:1281–1295
Rousseeuw P (1986) Multivariate estimation with high breakdown point. In: Grossman W, Pflug G, Vincze I, Wertz W (eds) Mathematical statistics and applications. Reidel, Dordrecht, pp 283–297
Rousseeuw P, Croux C, Todorov V, Ruckstuhl A, Salibian-Barrera M, Verbeke T, Koller M, Maechler M (2012) Robustbase: basic robust statistics. R package version 0.9-2
Ruiz-Gazen A, Berro A, Larabi Marie-Sainte S, (2010) Detecting multivariate outliers using projection pursuit with particle swarm optimization. Compstat 2010:89–98
Shaker AJ, Prendergast LA (2011) Iterative application of dimension reduction methods. Electron J Stat 5:1471–1494
Tibshirani R (2013) Bootstrap: functions for the book "An introduction to the bootstrap". R package version 2012.04-1
Tyler DE (1987) A distribution-free M-estimator of multivariate scatter. Ann Stat 15:234–251
Tyler DE, Critchley F, Dümbgen L, Oja H (2009) Invariant co-ordinate selection. J Roy Stat Soc 71:549–592
Venables WN, Ripley BD (2002) Modern applied statistics with S, 4th edn. Springer, New York
Weisberg S (2002) Dimension reduction regression in R. J Stat Softw 7:1–22
Ye Z, Weiss RE (2003) Using the bootstrap to select one of a new class of dimension reduction methods. J Am Stat Assoc 98:968–979
Zhou ZH (2012) Ensemble methods. CRC Press, Boca Raton, Foundations and Algorithms

On the Computation of Symmetrized M-Estimators of Scatter

Jari Miettinen, Klaus Nordhausen, Sara Taskinen and David E. Tyler

1 Introduction

Almost all of the classical multivariate methods, including principal component analysis, multivariate regression, canonical correlation analysis, etc., are dependent on the use of the sample covariance matrix. It is well-known that under the assumption of multivariate normality, the methods based on this estimator are optimal. However, if the normality assumption is not satisfied, e.g., if the data are contaminated with outlying observations or have heavier tails than that of the normal distribution, then methods based on the sample covariance matrix perform poorly.

A widely used approach for robustifying classical multivariate methods is the so-called plug-in approach. In this approach, the sample covariance matrix is replaced by a robust scatter matrix. As a consequence, a vast variety of robust alternatives for the sample covariance matrix have been proposed in the literature. Some widely used robust estimators include M-estimators (Maronna 1976; Huber 1981), MCD-estimators (Rousseeuw 1985), and S-estimators (Davies 1987; Lopuhaä 1989) among others. For an overview of robust multivariate methods, see Maronna et al. (2006).

J. Miettinen (✉) · S. Taskinen
Department of Mathematics and Statistics, University of Jyvaskyla, 40014 Jyvaskyla, Finland
e-mail: jari.p.miettinen@jyu.fi

S. Taskinen
e-mail: sara.l.taskinen@jyu.fi

K. Nordhausen
Department of Mathematics and Statistics, University of Turku, 20014 Turku, Finland
e-mail: klaus.nordhausen@utu.fi

K. Nordhausen
School of Health Sciences, University of Tampere, 30014 Tampere, Finland

D.E. Tyler
Department of Statistics, Rutgers University, New Brunswick, NJ 08854, USA
e-mail: dtyler@rci.rutgers.edu

C. Agostinelli et al. (eds.), *Recent Advances in Robust Statistics: Theory and Applications*, DOI 10.1007/978-81-322-3643-6_8

When robust plug-in methods are proposed, one important issue is often ignored, namely that a multivariate method may not be valid unless the robust scatter matrix satisfies certain crucial properties that hold for the sample covariance matrix. In Nordhausen and Tyler (2015) a thorough discussion of such properties is given. Focusing on the so-called joint and block independence properties (defined in the next section), Nordhausen and Tyler (2015) give several examples of plug-in multivariate methods, which are not valid unless the scatter matrix possesses these properties. Examples include independent component analysis, observational regression, and graphical modeling. For the role of scatter matrices in independent component analysis, see also Oja et al. (2006), Nordhausen et al. (2008), Tyler et al. (2009), among others.

In Oja et al. (2006), it is shown that by computing any scatter matrix using pairwise differences rather than the observations themselves produces an estimator with the joint independence property. Sirkiä et al. (2007) discuss general symmetrized M-estimators and give as examples the symmetrized Huber estimators, and Dümbgen's (1998) estimator, which is a symmetrized version of Tyler's (1987) M-estimator. Croux et al. (1994), Roelant et al. (2009) propose using symmetrized S-estimators in univariate and multivariate regression settings, respectively, with their main focus being on improving efficiency at the normal model.

As symmetrized estimators are defined using pairwise differences, the computations become intensive with increasing sample size. In this paper, we focus on the computational aspects and consider a few practical ways to handle this problem, especially in the context of M-estimates. The paper is organized as follows: In Sect. 2 we recall the definitions of scatter matrix and block and joint independence, and in Sect. 3 the definition and main properties of symmetrized M-estimators of scatter. Section 4 provides some new approaches for computing symmetrized estimators. In Sects. 5 and 6 simulation studies are given to compare efficiencies and computation times of different approaches, respectively. The paper is concluded with some discussion in Sect. 7.

2 Scatter Matrices and Block Independence

Recall first the definition of a scatter matrix functional.

Definition 1 Let $\mathbf{x}$ be a p-variate random vector with cumulative distribution function $F_{\mathbf{x}}$. Then a $p \times p$ matrix valued functional $\mathbf{V} = \mathbf{V}(F_{\mathbf{x}})$ is a *scatter matrix functional* if it is symmetric, positive semi-definite and affine equivariant in the sense that

$$\mathbf{V}(F_{\mathbf{Ax}+\mathbf{b}}) = \mathbf{A}\mathbf{V}(F_{\mathbf{x}})\mathbf{A}^t \quad (1)$$

for any full rank $p \times p$ matrix $\mathbf{A}$ and any p-vector $\mathbf{b}$.

A scatter matrix is then naturally defined as $\hat{\mathbf{V}} = \mathbf{V}(F_n)$, where F_n is the empirical cdf. Most robust counterparts of covariance matrix satisfy (1). However, they usually do not satisfy the so-called joint and block independence properties, which are characteristic of the covariance matrix and are defined as follows.

Definition 2 Assume that $\mathbf{x} = (\mathbf{x}_1^t, \ldots, \mathbf{x}_k^t)^t$ is a p-vector consisting of k mutually independent subvectors with dimension p_i, $i = 1, \ldots, k$, such that $\sum_{i=1}^k p_i = p$.

(i) The scatter matrix functional $\mathbf{V}(F_\mathbf{x})$ is said to have the *block independence property*, if it is a block diagonal matrix with block sizes p_i, $i = 1, \ldots, k$.
(ii) If $k = p$, which means that $\mathbf{x}$ has independent components, and $\mathbf{V}(F_\mathbf{x})$ is a diagonal matrix, then it is said to have the *joint independence property*.

Note that the block independence property implies the joint independence property, but not vice versa. In Nordhausen and Tyler (2015), several examples of multivariate methods are given for which it is necessary for a scatter matrix to possess the joint or block independence property.

Most scatter functionals do not posses the joint or block independence property. A common conjecture here is that only scatter matrices which can be expressed as functions of pairwise differences have this property. For example $\mathbf{COV}(\mathbf{x}) = E((\mathbf{x} - E(\mathbf{x}))(\mathbf{x} - E(\mathbf{x}))^t) = 2^{-1}E((\mathbf{x}_1 - \mathbf{x}_2)(\mathbf{x}_1 - \mathbf{x}_2)^t)$, where $\mathbf{x}_1$ and $\mathbf{x}_2$ denote independent copies of $\mathbf{x}$, can be written in such a way.

What about scatter matrices which cannot be expressed using pairwise differences? A quite simple but ingenious approach is to apply a scatter functional to the pairwise differences, which is known as symmetrization. Theorem 1 in Oja et al. (2006) shows that when a scatter matrix functional is applied to the pairwise differences of the observations, then the resulting functional possesses the joint independence property. In Nordhausen and Oja (2011), Nordhausen and Tyler (2015), it is shown that symmetrization yields to a more general block independence property.

A formal definition of symmetrization is given as follows.

Definition 3 Let $\mathbf{V}(F_\mathbf{x})$ be any scatter functional. Then the corresponding *symmetrized scatter functional* is defined as

$$\mathbf{V}_s(F_\mathbf{x}) = \mathbf{V}(F_{\mathbf{x}_1 - \mathbf{x}_2}),$$

where $\mathbf{x}_1$ and $\mathbf{x}_2$ are two independent copies of $\mathbf{x}$.

In this paper, we are mainly interested in computational aspects of symmetrized scatter matrices. Although the computational issues discussed herein apply to any symmetrized scatter matrix, this paper focuses on symmetrized M-estimators of scatter (Sirkiä et al. 2007), which, as to be seen, can be made computationally feasible for fairly large sample sizes. The next section reviews the definition and basic properties of symmetrized M-estimators of scatter.

3 Symmetrized M-Estimators of Scatter

Write again $\mathbf{x}$ for a p-variate random vector with cumulative distribution function $F_{\mathbf{x}}$. In this paper, we focus on elliptically symmetric distributions. Such a distribution family is often used in robustness studies as it includes distributions with heavy tails (e.g., elliptical Cauchy distribution) as well as distributions which can be used to generate atypical observations (e.g. contaminated normal distribution).

An elliptically symmetric distribution is obtained as an affine transformation of a spherical distribution. Recall that a p-variate random vector $\mathbf{z}$ is spherically symmetric around the origin if $\mathbf{Uz} \sim \mathbf{z}$ for all orthogonal $p \times p$ matrices $\mathbf{U}$. Then $\mathbf{x} = \boldsymbol{\Omega}\mathbf{z} + \boldsymbol{\mu}$, where $\boldsymbol{\Omega}$ is a full rank $p \times p$ matrix and $\boldsymbol{\mu}$ a p-vector, has an elliptically symmetric distribution with density of the form

$$f(\mathbf{x}; \boldsymbol{\mu}, \boldsymbol{\Sigma}, g) = |\boldsymbol{\Sigma}|^{-1/2} g(\boldsymbol{\Sigma}^{-1/2}(\mathbf{x} - \boldsymbol{\mu})),$$

where $g(\mathbf{z}) = \exp(-\rho(||\mathbf{z}||))$ represents the density of $\mathbf{z}$, with $\rho(\cdot)$ being a nonnegative function and $\boldsymbol{\Sigma} = \boldsymbol{\Omega}\boldsymbol{\Omega}^t$. Without loss of generality, $\boldsymbol{\Sigma}^{1/2}$ is taken to be the symmetric positive definite square root of $\boldsymbol{\Sigma}$. Note that the density of $\mathbf{z}$ depends only on the value of its radius $||\mathbf{z}||$, and the function $\rho(\cdot)$ does not depend on the parameters $\boldsymbol{\mu}$ and $\boldsymbol{\Sigma}$.

The parameter $\boldsymbol{\mu}$ is the location center of the distribution and the scatter matrix $\boldsymbol{\Sigma}$ is proportional to the regular covariance matrix (if it exists). Examples of function $g(\cdot)$ include $g(\mathbf{z}) = (2\pi)^{-p/2} \exp(-\mathbf{z}^t\mathbf{z}/2)$, which corresponds to the p-variate normal distribution and

$$g(\mathbf{z}) = \frac{\Gamma((p+\nu)/2)}{\Gamma(\nu/2)(\pi\nu)^{p/2}} \left(1 + \frac{\mathbf{z}^t\mathbf{z}}{\nu}\right)^{-(p+\nu)/2},$$

which corresponds to the p-variate t-distribution on ν degrees of freedom. Within the class of elliptical distribution, i.e., for unknown g, only the location $\boldsymbol{\mu}$ and the "shape" of $\boldsymbol{\Sigma}$, i.e., the value of $\boldsymbol{\Sigma}$ up to proportionality, is well defined, whereas the constant of proportionality is confounded with the function g.

Next, we recall the definition of the symmetrized M-functional as given in Sirkiä et al. (2007).

Definition 4 Assume that $\mathbf{x}$ is a p-variate random vector with cdf $F_{\mathbf{x}}$, and let $\mathbf{x}_1$ and $\mathbf{x}_2$ be two independent copies of $\mathbf{x}$. A *symmetrized M-functional* $\mathbf{V}_s = \mathbf{V}_s(F_{\mathbf{x}_1-\mathbf{x}_2})$ is defined as a solution to

$$E[w_1(r_{12})(\mathbf{x}_1 - \mathbf{x}_2)(\mathbf{x}_1 - \mathbf{x}_2)^t - w_2(r_{12})\mathbf{V}_s] = \mathbf{0},$$

where $r_{12} = [(\mathbf{x}_1 - \mathbf{x}_2)^t\mathbf{V}_s^{-1}(\mathbf{x}_1 - \mathbf{x}_2)]^{1/2}$, and w_1 and w_2 are some real-valued functions on $[0, \infty)$.

Sirkiä et al. (2007) observe that the assumptions on the weight functions and on the distribution of the pairwise differences needed for the existence and uniqueness of symmetrized M-functionals follow from Huber's (1981) results for (non-symmetrized) M-functionals. When $\mathbf{x}$ has an elliptical distribution, $\mathbf{V}_s \propto \boldsymbol{\Sigma}$, with the constant of proportionality being dependent on the weight functions w_1 and w_2 and the density g, but not on the parameters $\boldsymbol{\mu}$ or $\boldsymbol{\Sigma}$.

An estimator corresponding to a scatter matrix functional $\mathbf{V}_s$ is obtained when $F_{\mathbf{x}_1-\mathbf{x}_2}$ in Definition 4 is replaced with the empirical distribution function of the pairwise differences. A symmetrized M-estimator of scatter, $\hat{\mathbf{V}}_s$, then solves

$$\binom{n}{2}^{-1}\sum\sum_{i<j}[w_1(r_{ij})(\mathbf{x}_i-\mathbf{x}_j)(\mathbf{x}_i-\mathbf{x}_j)^t - w_2(r_{ij})\mathbf{V}_s] = \mathbf{0},$$

where w_1 and w_2 are real-valued functions on $[0,\infty)$. Notice that choices $w_1(r) = \rho'(r)/r$ and $w_2(r) = 2$ yield the maximum likelihood estimator under a specific elliptical distribution. The robustness properties and limiting distributions of general symmetrized M-estimators were discussed in Sirkiä et al. (2007).

In this paper, we consider the following symmetrized M-estimators:

- The sample covariance matrix, which corresponds to $w_1(r) = 1$ and $w_2(r) = 2$, or equivalently to $w_1(r) = 1/2$ and $w_2(r) = 1$
- The symmetrized Cauchy M-estimator, which has weight functions corresponding to those of the maximum likelihood estimator for the elliptical Cauchy distribution, i.e., $w_1(r) = (1+p)/(1+r^2)$ and $w_2(r) = 1$. It is worth noting that this is not the same as the maximum likelihood estimator based on the pairwise differences from a random sample of an elliptical Cauchy distribution.
- The symmetrized Huber estimators, which have weight functions $w_2(r) = 1$ and

$$w_1(r) = \begin{cases} 1/\sigma^2, & r^2 \le c^2 \\ c^2/(r^2\sigma^2), & r^2 > c^2, \end{cases}$$

  where c is a tuning constant defined so that $q = Pr(\chi_p^2 \le c^2/2)$ for a chosen q. The scaling factor σ is chosen so that $E[w_1(||\mathbf{x}_1 - \mathbf{x}_2||)] = p$, where $\mathbf{x}_1, \mathbf{x}_2 \sim N(\mathbf{0}, \mathbf{I}_p)$, which makes the estimator Fisher-consistent for $\boldsymbol{\Sigma}$ at the multivariate normal model.
- Dümbgen's (1998) estimator, which corresponds to choosing $w_1(r) = p/r^2$ and $w_2(r) = 1$.

Dümbgen's estimator is only defined up to proportionality, i.e., both $\hat{\mathbf{V}}_{s,1}$ and $\hat{\mathbf{V}}_{s,2}$ satisfy the corresponding estimating equations, if and only if $\hat{\mathbf{V}}_{s,1} \propto \hat{\mathbf{V}}_{s,2}$. Furthermore, as noted previously, under sampling from an elliptical distribution, the symmetrized Cauchy M-estimator is Fisher consistent for the parameter $\boldsymbol{\Sigma}$ only up to proportionality. This is also true of the sample covariance matrix, and the symmetrized Huber

M-estimator at elliptical models other than the multivariate normal. These factors, though, are not important to the efficiency comparisons given in Sect. 5, since only the shape of the scatter matrices are considered in these comparisons.

4 Computation of Symmetrized M-Estimators

Hereafter, we consider only the case $w_2(\cdot) = 1$, which agrees with the original definition of the M-estimators given in Maronna (1976). Note that this case holds for the three M-estimators discussed in the previous section, as well as for the maximum likelihood estimators of scatter under an elliptical family of distributions, i.e., with a fixed g. A general recent overview of the M-estimators of scatter for the case $w_2(\cdot) = 1$ can be found in Dümbgen et al. (2015). They point out that the most commonly used method to compute such M-estimates is via a simple fixed point algorithm, which is known to converge under very general conditions to a unique solution, regardless of the initial value, as shown in Kent and Tyler (1991).

Assume in the following that we have a sample of n vectors $\mathbf{X} = (\mathbf{x}_1, \ldots, \mathbf{x}_n)$ and the goal is to compute the symmetrized M-scatter matrix $\mathbf{V}_s$ of interest. The most naive approach would be to apply the fixed point algorithm for the unsymmetrized scatter of interest to all $n(n-1)$ pairwise differences $\mathbf{x}_i - \mathbf{x}_j$ with $i \neq j$. Notice that now the location center does not need to be estimated as for the symmetrized vectors the location center is naturally the origin. Nevertheless, even for a moderate sample size n, the computational burden can be tremendous and so new approaches are needed to to deal with this. In the following, we will consider a few practical ways to reduce the computational burden and memory demand.

The number of pairwise differences can be halved since only the $n(n-1)/2$ pairwise differences $\mathbf{x}_i - \mathbf{x}_j$ with $i < j$ are needed to compute the symmetrized scatter matrix. Hence the most basic algorithm is *the fixed point algorithm* with updating step at iteration k:

$$\mathbf{V}_s^{k+1}(\mathbf{X}) = \binom{n}{2}^{-1} \sum_{i<j} \left\{ w_1(r_{ij}^k)(\mathbf{x}_i - \mathbf{x}_j)(\mathbf{x}_i - \mathbf{x}_j)^t \right\},$$

where r_{ij}^k is based on the current scatter estimate $\mathbf{V}_s^k$. Recently, Nordhausen and Tyler (2015) suggested rewriting the above algorithm as

$$\mathbf{V}_s^{k+1}(\mathbf{X}) = 2(n(n-1))^{-1} \sum_{i=2}^{n} \mathbf{S}_i^{k+1}(\mathbf{X}),$$

where

$$\mathbf{S}_i^{k+1}(\mathbf{X}) = \sum_{j=1}^{i-1} w_1(r_{ij}^k)(\mathbf{x}_i - \mathbf{x}_j)(\mathbf{x}_i - \mathbf{x}_j)^t.$$

The computation of $\mathbf{S}_i^{k+1}(\mathbf{X})$ then can be naturally divided into several threads. We refer to this algorithm as *the parallel algorithm* and study, in Sect. 6, how much this approach speeds up computations.

Dümbgen et al. (2016) have recently argued that using a fixed point algorithm for computing M-estimates of scatter can be less than optimal. They consider several alternative algorithms and recommended *a partial Newton (PN) algorithm*, which, in most cases, is considerably faster. The basic idea behind the PN algorithm is to first perform a few fixed point steps and to then evaluate whether shifting to a Newton–Raphson step with an approximated Hessian is better. We refer to the reader to the aforementioned paper for more details regarding the algorithm. A restriction of the PN algorithm is that the weight functions must be smooth, which excludes, for example, Huber's weight functions. Two versions of the PN algorithm were introduced in Dümbgen et al. (2016), with one version requiring all pairwise differences $\mathbf{x}_i - \mathbf{x}_j$ with $i < j$ being in the memory, and the other version being a sequential algorithm which avoids storing all pairwise differences. The sequential algorithm seems to be, in most cases, faster than the one that stores all pairwise differences.

We have, thus, several algorithms available so far for the computation of symmetrized M-estimators of scatter. However, these are all computationally intensive as they either store all pairwise differences $\mathbf{x}_i - \mathbf{x}_j$ with $i < j$ in the memory or compute sequentially all quantities of interest. This computational burden is demonstrated later in Sect. 6.

A possible way to ease this computational problem can be motivated by noting the resemblance of the symmetrized scatter matrix to a U-statistic of order two. Recall that for a sample $\mathbf{X} = (\mathbf{x}_1, \ldots, \mathbf{x}_n)$ a U-statistic for a parameter θ based on a symmetric kernel $h(\mathbf{x}_{(1)}, \ldots, \mathbf{x}_{(K)})$ of order K is defined as

$$U = N^{-1} \sum_{i=1}^{N} h(\mathbf{x}_{(1)}, \ldots, \mathbf{x}_{(K)}),$$

where $N = \binom{n}{K}$ and the kernel is computed for all possible subsamples of size K denoted by $\mathbf{x}_{(1)}, \ldots, \mathbf{x}_{(K)}$. A simple example of a U-statistic is the sample covariance matrix, which has a kernel of order two and can be expressed as

$$h(\mathbf{x}_{(1)}, \mathbf{x}_{(2)}) = 2^{-1}(\mathbf{x}_{(1)} - \mathbf{x}_{(2)})(\mathbf{x}_{(1)} - \mathbf{x}_{(2)})^t$$

and hence

$$\mathbf{COV}(\mathbf{X}) = \binom{n}{2}^{-1} \sum_{i \neq j} 2^{-1}(\mathbf{x}_{(i)} - \mathbf{x}_{(j)})(\mathbf{x}_{(i)} - \mathbf{x}_{(j)})^t.$$

In general, though, not all symmetrized scatter matrices can be expressed as U-statistics, since they typically have only an implicit rather than an explicit representation in terms of pairwise differences.

In the context of U-statistics, Blom (1976) noted that it is possible to use less than N terms without losing much information when estimating θ, and he called such estimates incomplete U-statistics. Such estimates have also been referred to as weighted U-statistics, with weights 0 or 1, or as reduced U-statistics. In Blom (1976), Brown and Kildea (1978), the statistical properties of incomplete U-statistics are derived.

Following the idea of incomplete U-statistics, many ways to choose the terms used in computations are possible. The most basic one is independent subsampling, where m sets out of all N sets are chosen at random. This can, however, give different weight to different observations in the data. Another convenient choice for kernels of order $K = 2$, which gives each observation equal weight, is what we refer to as a "running average of length m." For this purpose, we treat the ordering of the data as cyclic and define an extended data matrix $\mathbf{X}^* = (\mathbf{x}_1^*, \ldots, \mathbf{x}_{n+m}^*) = (\mathbf{x}_1, \ldots, \mathbf{x}_n, \mathbf{x}_1, \ldots, \mathbf{x}_m)$. Our *incomplete symmetrized M-estimator of length m*, $\hat{\mathbf{V}}_I$, then solves

$$\mathbf{V}_I = \frac{1}{nm} \sum_{i=1}^{n} \sum_{j=i+1}^{i+m} w_1(r_{ij})(\mathbf{x}_i^* - \mathbf{x}_j^*)(\mathbf{x}_i^* - \mathbf{x}_j^*)^t.$$

In the following, we explore the idea of computing symmetrized scatter matrices by using running averages of different lengths m and compare the loss in efficiency to the gain in computation time. From a practical point of view, the observation order should be randomly permuted in order to avoid the effect of how the data was recorded. Using permutations in the simulations, though, are not needed since the simulated data set follows the same model as any permutation of it.

5 Efficiency Comparisons

In this section, we compare the efficiencies of the incomplete (using only mn pairwise differences) symmetrized estimators to that of the corresponding complete symmetrized estimator. We include the symmetrized Huber estimator with $q = 0.90$, Dümbgen's estimator and the symmetrized Cauchy M-estimator in the comparisons. Since, as previously discussed, the estimators are only comparable up to proportionality, we standardize all estimators so that their traces are equal to p.

To compare the finite sample efficiencies, we first carried out a simulation study with samples of size $n = 1000$, 2000, and 4000, dimensions $p = 3$ and $p = 8$, and under the normal distribution (N), the contaminated normal distribution (cN), and the t-distribution on five degrees of freedom (t_5). The cumulative distribution function of the contaminated normal distribution is $\Phi_{\varepsilon,c}(\mathbf{x}) = (1-\varepsilon)\Phi(\mathbf{x}) + \varepsilon\Phi(\mathbf{x}/c)$, where $\varepsilon, c > 0$ and Φ denotes the cumulative distribution function of the standard

Table 1 Asymptotic efficiencies of the symmetrized Huber M-estimator, Dümbgen's estimator and the symmetrized Cauchy M-estimator relative to the sample covariance matrix. The asymptotic relative efficiencies are evaluated at the normal (N), the contaminated normal (cN), and t-distribution on five degrees of freedom (t_5)

	p = 3			p = 8		
	N	cN	t_5	N	cN	t_5
Symmetrized Huber	0.99	1.67	2.12	1.00	1.65	2.12
Dümbgen	0.93	2.27	2.40	0.96	2.43	2.57
Symmetrized Cauchy	0.94	2.20	2.38	0.96	2.40	2.54

multivariate normal distribution. In our simulation settings, we used $\varepsilon = 0.1$ and $c = 3$.

The asymptotic efficiencies of the three standardized robust scatter estimators relative to the standardized sample covariance matrix are listed in Table 1. The asymptotic relative efficiencies were computed using the results in Sirkiä et al. (2007), wherein they observed that the symmetrized Huber estimator and the Dümbgen's estimator are highly efficient not only at heavy tailed distributions but also at the multivariate normal distribution. The symmetrized Cauchy M-estimator, though, suffers from some efficiency loss at the multivariate normal distribution case.

To compare the finite sample efficiencies, the mean squared errors of the off-diagonal elements of the standardized scatter matrices, that is,

$$MSE(\hat{\mathbf{V}}) = \frac{2}{Np(p-1)} \sum_{k=1}^{N} \sum_{i=1}^{p-1} \sum_{j=i+1}^{p} (\hat{\mathbf{V}}_{ij}^{(k)} - \mathbf{I}_{ij})^2,$$

were computed using $N = 2000$ samples. The efficiencies were then defined by taking the ratios of the corresponding MSEs. The results are listed in Tables 2, 3 and 4. For all of the estimators, there is some loss, but somewhat surprising not a large loss, in efficiency when $m = 10$, and when $m = 20$, the efficiency loss is always less than 5 %. The loss in efficiency is slightly worst for the Dümbgen's estimator than for the other estimators.

Among the symmetrized scatter matrices considered in this paper, only the sample covariance matrix is a U-statistic. Nevertheless, the simulations indicate that all scatter matrices computed using running averages of length m seem to behave in a similar fashion. These empirical results suggest that theoretical results obtained for the incomplete sample covariance matrix may give us insight into the behavior of other incomplete symmetrized estimates of scatter.

In particular, results from Brown and Kildea (1978) for incomplete U-statistics allow us to compute the asymptotic relative efficiency of the incomplete sample covariance estimator with respect to the complete sample covariance matrix. For a spherically symmetric distribution with $\mathbf{COV}(\mathbf{z}) = \mathbf{I}_p$, the efficiency of the incomplete symmetrized sample covariance matrix relative to the complete one is

Table 2 Finite sample relative efficiencies (MSE from 2000 samples) of the incomplete symmetrized Huber M-estimator with respect to the complete estimator

	m	p = 3			p = 8		
		N	cN	t_5	N	cN	t_5
n = 1000	10	0.94	0.95	0.94	0.95	0.96	0.95
	20	0.97	0.97	0.98	0.97	0.98	0.97
	50	0.98	0.98	0.98	0.98	0.98	0.98
	100	0.97	0.99	0.98	0.98	0.98	0.98
n = 2000	10	0.94	0.95	0.95	0.95	0.96	0.95
	20	0.97	0.98	0.97	0.97	0.98	0.97
	50	0.99	0.99	0.98	0.98	0.99	0.99
	100	0.99	0.99	0.99	0.99	0.99	0.99
n = 4000	10	0.95	0.97	0.96	0.95	0.96	0.97
	20	0.98	0.98	0.97	0.97	0.98	0.98
	50	0.99	1.00	0.99	0.99	0.99	0.99
	100	0.99	0.99	0.99	0.99	0.99	0.99

Table 3 Finite sample relative efficiencies (MSE from 2000 samples) of incomplete Dümbgen's estimators with respect to the complete estimator

	m	p = 3			p = 8		
		N	cN	t_5	N	cN	t_5
n = 1000	10	0.90	0.93	0.91	0.94	0.94	0.94
	20	0.95	0.95	0.96	0.96	0.97	0.97
	50	0.97	0.97	0.97	0.98	0.98	0.98
	100	0.98	0.97	0.97	0.98	0.98	0.98
n = 2000	10	0.90	0.92	0.92	0.93	0.94	0.94
	20	0.94	0.96	0.96	0.97	0.97	0.97
	50	0.98	0.98	0.98	0.98	0.98	0.99
	100	0.98	0.99	0.98	0.99	0.99	0.99
n = 4000	10	0.90	0.92	0.92	0.94	0.94	0.95
	20	0.95	0.96	0.96	0.97	0.97	0.97
	50	0.98	0.97	0.99	0.98	0.99	0.99
	100	0.98	0.99	0.99	0.99	0.99	0.99

$$ARE(\hat{\mathbf{V}}_s, \hat{\mathbf{V}}_I^{(m)}) = \frac{2m\kappa}{0.5 + (2m-1)\kappa}, \tag{2}$$

where $\kappa = E[z_i^2 z_j^2]/(2(E[z_i^2 z_j^2] + 1))$, and z_i and z_j are different components of $\mathbf{z}$. In Fig. 1 we plot the asymptotic relative efficiency of the symmetrized incomplete sample covariance matrix as a function of m for the 3-variate normal, contaminated normal and t_5-distribution, for which κ = 1/4, 25/68, and 3/8 respectively. We also

Table 4 Finite sample relative efficiencies (MSE from 2000 samples) of incomplete Cauchy M-estimators with respect to the complete estimator

	m	p = 3			p = 8		
		N	cN	t_5	N	cN	t_5
n = 1000	10	0.93	0.94	0.95	0.94	0.95	0.95
	20	0.96	0.97	0.98	0.97	0.97	0.97
	50	0.98	0.98	0.98	0.98	0.98	0.98
	100	0.98	0.99	0.98	0.98	0.98	0.98
n = 2000	10	0.93	0.95	0.95	0.94	0.95	0.95
	20	0.96	0.97	0.98	0.97	0.97	0.97
	50	0.98	0.98	0.98	0.98	0.99	0.99
	100	0.98	0.99	0.99	0.99	0.99	0.99
n = 6000	10	0.93	0.94	0.93	0.94	0.95	0.95
	20	0.96	0.97	0.96	0.97	0.97	0.97
	50	0.98	0.99	0.99	0.99	0.99	0.99
	100	0.98	0.99	0.99	0.99	0.99	0.99

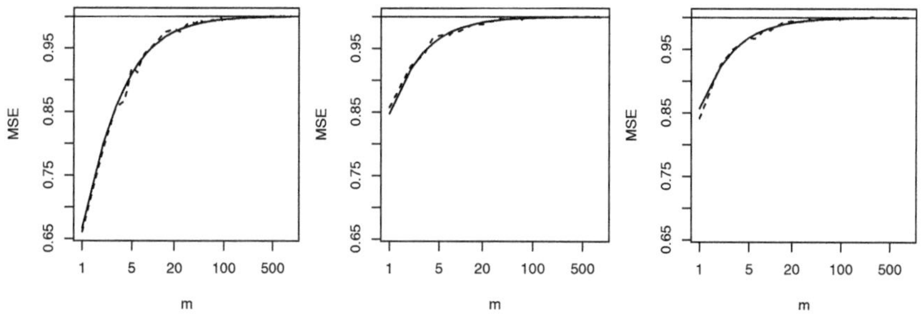

Fig. 1 Finite sample efficiencies of the incomplete symmetrized sample covariance matrix with respect to the symmetrized sample covariance matrix (*dashed lines*) for different distributions with $n = 1000$ and $p = 3$. The distributions from *left* to *right* are the normal distribution, the contaminated normal distribution and the t_5-distribution. The *solid lines* give the asymptotic relative efficiencies

simulated the finite sample efficiencies, which correspond to the dash lines in the figures, computed as ratios of MSEs using $n = 1000$ and $N = 2000$ repetitions. It can be seen that the efficiencies increase rapidly as a function of m, with a limit of one as $m \to \infty$. Interestingly, the efficiency at $m = 1$ is notably higher in the case of heavy tailed distributions than in the case of the normal distribution. The choice $m = 20$ is sufficient to produce an estimator with very high efficiency.

All simulations so far have focused on data coming from an elliptical model, among which the only distribution with independent marginals is the multivariate normal distribution. However, there are many areas of applications, such as independent components analysis, for which independent marginals outside of the multivariate normal distribution are of interest. Consequently, we also simulated data for different sample sizes and dimensions from a model with mutually independent components, where each component has a standard exponential distribution. Here, if the scatter functional possesses the joint independence property, then the off-diagonal values of the scatter matrix are equal to zero. For this setting, we compare the symmetrized M-estimators (Dümbgen's estimator and the Cauchy M-estimator) based on all pairwise differences to the corresponding estimators using running averages of length 20. Figure 2 gives the mean squared errors of the off-diagonal values based on 1000 repetitions. The figure shows that in this case the incomplete estimators with $m = 20$ behave similarly to the regular symmetrized estimators.

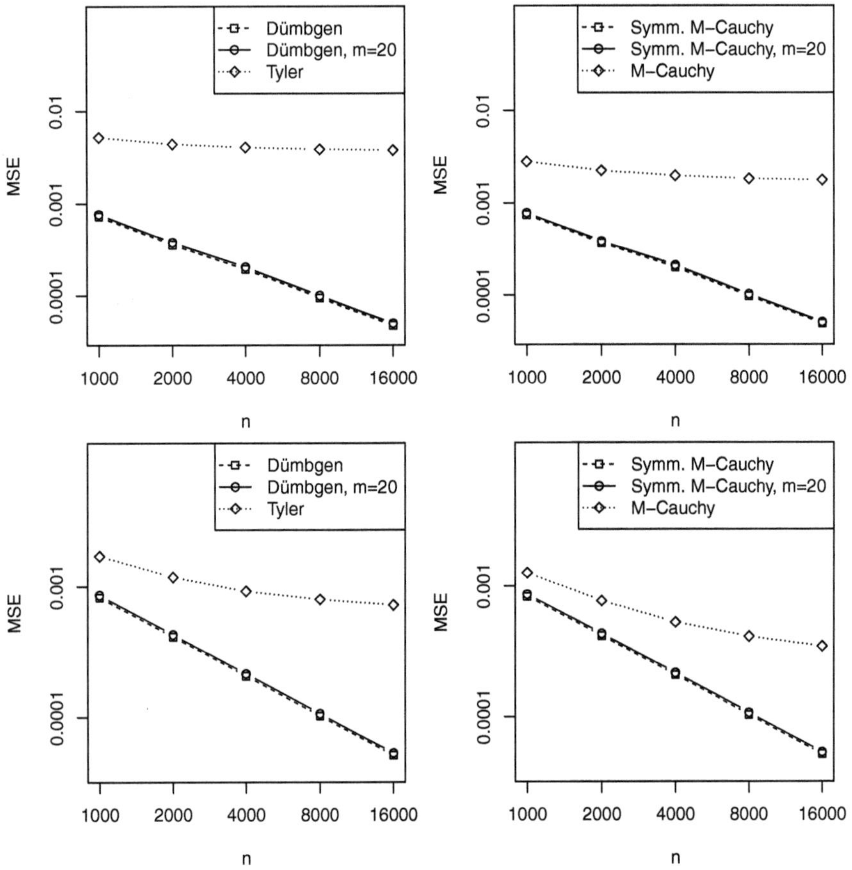

Fig. 2 Mean squared errors of the off-diagonal elements of the Dümbgen's estimate, the incomplete Dümbgen's estimate with $m = 20$, and Tyler's estimate on the *left*, and MSE of the symmetrized Cauchy M-estimate, the incomplete symmetrized Cauchy M-estimate with $m = 20$, and nonsymmetrized Cauchy M-estimate on the *right*, when the three-dimensional (*on the top row*) and eight-dimensional (*on the bottom row*) data are generated from a distribution with mutually independent and exponentially (with mean 1) distributed components

For comparison, we also compute the corresponding non-symmetrized versions of the scatter matrices (Tyler's estimator and the Cauchy M-estimator, which corresponds to the MLE for the Cauchy distribution, respectively). These estimators do not possess the desired joint independence property, and so the corresponding functionals of these nonsymmetrized versions do not have zero off-diagonal elements even though the variables are independent. Consequently, as seen in Fig. 2, their MSEs do not go to zero as n increases.

As this section demonstrates, using running average sets of pairwise differences with small to moderate values of m results in only a small loss of efficiency relative to their complete version. In the next section, we will see how this small loss in efficiency pays off in computation time.

6 Computation Times

In comparing the computation times of the different algorithms presented in Sect. 4, we again chose the two dimensions $p = 3$ and $p = 8$, and use five sample sizes $n = 1000, 2000, 4000, 8000, 16{,}000$. For each combination of p and n, 50 samples from the multivariate t-distribution with five degrees of freedom are generated, and the computation times of the different algorithms are measured. The scatter matrices under consideration are the same as those used in previous sections. For the symmetrized Cauchy M-estimator and for the Dümbgen's estimator, both the fixed point algorithms and the partial Newton algorithms can be found in the R-packages ICSNP (Nordhausen et al. 2012) and fastM (Dümbgen et al. 2014), respectively. The symmetrized Huber estimator can also be computed using the R-package ICSNP. Currently, there are plans to implement the running average versions of the estimators in these packages.

Our main interest in the following comparisons is twofold. First, we are interested in when parallelization is beneficial, and second, in how fast the running average versions are relative to the standard implementations. In the simulations, we chose $m = 20$ for the incomplete estimators as this was in all cases considered to yield highly efficient estimators. We also used the partial Newton algorithm from the fastM package that uses sequential computations as this seems to be faster than having all pairwise differences in the memory (Dümbgen et al. 2016). All functions are mainly written in C or C++ with an R interface and should be therefore comparable (but have sometimes slightly different convergence criteria). The comparisons were done using R 3.1.1 (R Core Team 2014) on a Intel(R) Core(TM) i7-3770 CPU with 3.40 GHz, 32 GB of memory using 64-bit Red Hat Linux.

Medians of the computation times (on the logscale) of the symmetrized Cauchy M-estimator, Dümbgen's estimator and the symmetrized Huber estimator are given in Figs. 3, 4 and 5, respectively.

As expected, the regular fixed point algorithm utilizing all pairwise differences is slowest while the incomplete estimator is the fastest to compute. The ratio of their computation times is approximately the ratio of the number of pairs, which is

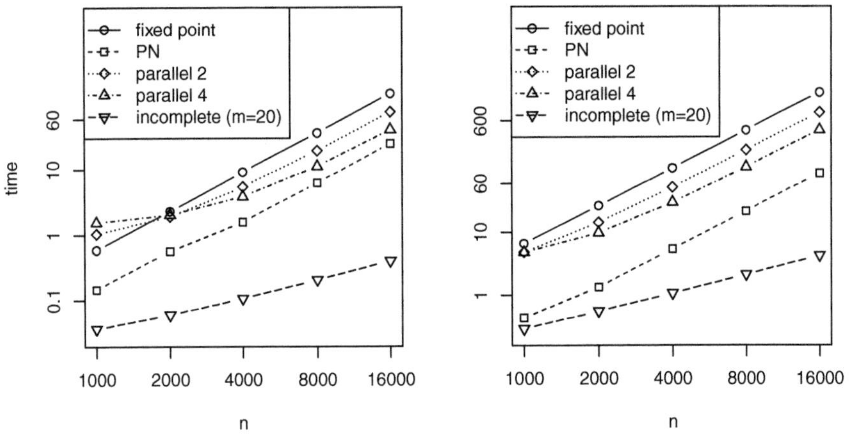

Fig. 3 Median computation times in seconds on a logscale for various algorithms used to compute the symmetrized Cauchy M-estimator. For each sample size, the median computation time is based on 50 independent random samples from the multivariate t_5-distribution. In the *left panel* $p = 3$ and in the *right panel* $p = 8$

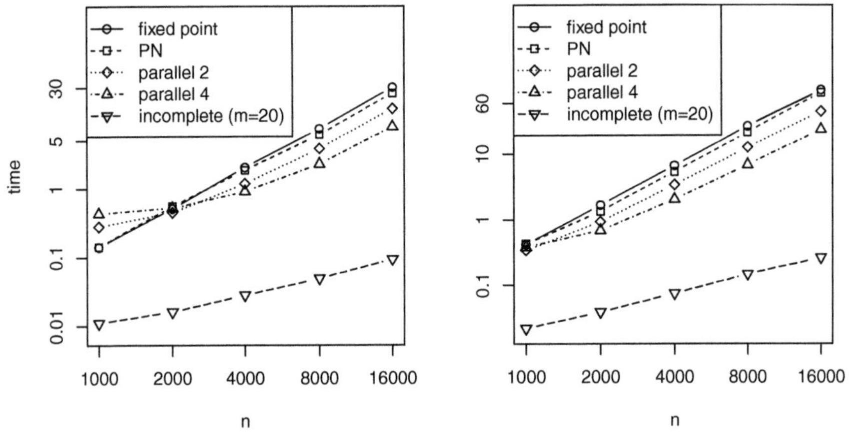

Fig. 4 Median computation times in seconds on a logscale for various algorithms used to compute Dümbgen's estimator. For each sample size, the median computation time is based on 50 independent random samples from the multivariate t_5-distribution. In the *left panel* $p = 3$ and in the *right panel* $p = 8$

$0.5(n-1)/m$. With large sample sizes, using two cores gains approximately 50 % in computation time and using four cores approximately 75 % relative to using only one core. We compared also the computation times when using six cores, but the computation times did not differ significantly different the version using four cores. Notice that the gain percentage of the parallel computation grows with the sample size

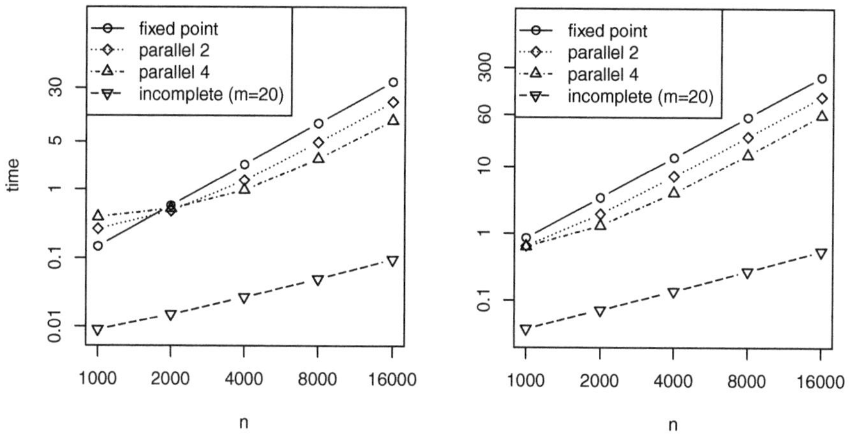

Fig. 5 Median computation times in seconds on a logscale for various algorithms used to compute the symmetrized Huber M-estimator with $q = 0.90$. For each sample size, the median computation time is based on 50 independent random samples from the multivariate t_5-distribution. In the *left panel* $p = 3$ and in the *right panel* $p = 8$

and the dimension until it reaches a limiting level. Parallelization becomes beneficial somewhere around $n = 2000$ when $p = 3$ and around $n = 1000$ when $p = 8$.

As already pointed out in Dümbgen et al. (2016), the partial Newton algorithm is not considerably faster than the fixed point algorithm when computing Dümbgen's estimator. However, the PN algorithm computes Dümbgen's estimator and the symmetrized Cauchy M-estimator equally fast, whereas all the other algorithms compute Dümbgen's estimator much faster than the symmetrized Cauchy M-estimator; for $p = 3$ and $p = 8$, approximately 5 and 18 times faster, respectively. Hence, the PN algorithm is superior to parallelized fixed point algorithm using four cores for the symmetrized Cauchy M-estimator, and vice versa for Dümbgen's estimator. Recall that the PN algorithm cannot be applied to the Huber estimator since the weight functions are not smooth. The computation time of the symmetrized Huber estimator is approximately the same as that of the Dümbgen's estimator when $p = 3$, but twice as long when $p = 8$.

7 Discussion

The relevance of symmetrized scatter matrices has only recently been recognized within the statistics literature. The benefit of using such scatter matrices is twofold: (i) they do not require a location estimate, and (ii) they possess the joint and the block independence properties, which are necessary properties for many multivariate methods. These benefits, however, come at a cost, namely that symmetrized scatter matrices tend to be more computationally intensive and are slightly less robust than their unsymmetrized counterparts. In this paper, the computational aspects of symmetrized M-estimators have been considered. In particular, it is shown that parallelization of the fixed-point algorithm is possible for these M-estimators and that this provides a considerable gain when, for example, four cores are used. Parallelization of the fixed point algorithm alone, though, is not as computationally efficient as more recently proposed partial Newton algorithms. Another computational alternative, proposed within the paper, is motivated by results on incomplete U-statistics, namely to reduce the number of pairwise differences used in computations. Such an approach proves to be promising. A huge gain in computation time is achieved with only a small loss in efficiency. Finally, we note that while the parallelization approach is specific for M-estimators, the subsampling of pairwise differences can be applied to any symmetrized scatter matrix.

Acknowledgments This work was supported by the Academy of Finland under Grants 251965, 256291, and 268703. David Tyler's work for this material was supported by the National Science Foundation under Grant No. DMS-1407751. We thank Dr. Seija Sirkiä for providing us the asymptotic relative efficiencies of the symmetrized estimators. The authors are grateful to the reviewers for their helpful comments.

References

Blom G (1976) Some properties of incomplete U-statistics. Biometrika 63(3):573–580

Brown BM, Kildea DG (1978) Reduced U-statistics and the Hodges-Lehmann estimator. Ann Stat 6(4):828–835. doi:10.1214/aos/1176344256

Croux C, Rousseeuw PJ, Hössjer O (1994) Generalized S-estimators. J Am Stat Assoc 89:1271–1278

Davies PL (1987) Asymptotic behaviour of S-estimates of multivariate location parameters and dispersion matrices. Ann Stat 15:1269–1292

Dümbgen L (1998) On Tyler's M-functional of scatter in high dimension. Ann Inst Stat Math 50:471–491

Dümbgen L, Nordhausen K, Schuhmacher H (2014) fastM: Fast computation of multivariate M-estimators. http://CRAN.R-project.org/package=fastM, R package version 0.0-2

Dümbgen L, Pauly M, Schweizer T (2015) M-functionals of multivariate scatter. Stat Surv 9:31–105

Dümbgen L, Nordhausen K, Schuhmacher H (2016) New algorithms for M-estimation of multivariate location and scatter. J Multivar Anal 144:200–217

Huber PJ (1981) Robust statistics. Wiley, New York

Kent JT, Tyler DE (1991) Redescending M-estimates of multivariate location and scatter. Ann Stat 19(4):2102–2119

Lopuhaä H (1989) On the relation between S-estimators and M-estimators of multivariate location and covariance. Ann Stat 17:1662–1683

Maronna RA (1976) Robust M-estimators of multivariate location and scatter. Ann Stat 4:51–67

Maronna RA, Martin DR, Yohai VJ (2006) Robust statistics: theory and methods. Wiley, Wiley Series in Probability and Statistics

Nordhausen K, Oja H (2011) Scatter matrices with independent block property and ISA. In: Proceedings of 19th European Signal Processing Conference 2011 (EUSIPCO 2011), pp 1738–1742

Nordhausen K, Tyler DE (2015) A cautionary note on robust covariance plug-in methods. Biometrika 102:573–588

Nordhausen K, Oja H, Ollila E (2008) Robust independent component analysis based on two scatter matrices. Austrian J Stat 37:91–100

Nordhausen K, Sirkiä S, Oja H, Tyler DE (2012) ICSNP: tools for multivariate nonparametrics. http://CRAN.R-project.org/package=ICSNP, R package version 1.0-9

Oja H, Sirkiä S, Eriksson J (2006) Scatter matrices and independent component analysis. Austrian J Stat 35:175–189

R Core Team (2014) R: a language and environment for statistical computing. R Foundation for Statistical Computing, Vienna, Austria. http://www.R-project.org/

Roelant E, Van Aelst S, Croux C (2009) Multivariate generalized S-estimators. J Multivar Anal 100:876–887

Rousseeuw PJ (1985) Multivariate estimation with high breakdown point. In: Grossmann W, Pflug G, Vincze I, Wertz W (eds) Mathematical methods and applications, vol B. Reidel, Dordrecht, Netherlands, pp 283–297

Sirkiä S, Taskinen S, Oja H (2007) Symmetrised M-estimators of scatter. J Multivar Anal 98:1611–1629

Tyler DE (1987) A distribution-free M-estimator of multivariate scatter. Ann Stat 15:234–251

Tyler DE, Critchley F, Dümbgen L, Oja H (2009) Invariant co-ordinate selection. J Roy Stat Soc Series B 71:549–595

Mortality and Life Expectancy Forecasting for a Group of Populations in Developed Countries: A Robust Multilevel Functional Data Method

Han Lin Shang

1 Introduction

Many statistical methods have been proposed for forecasting age-specific mortality rates (see Booth 2006; Booth and Tickle 2008; Shang et al. 2011; Tickle and Booth 2014, for reviews). Of these, a significant milestone in demographic forecasting was the work by Lee and Carter (1992). They applied a principal component method to age-specific mortality rates and extracted a single time-varying index of the level of mortality rates, from which the forecasts are obtained by a random-walk with drift. The method has since been extended and modified. For example, Renshaw and Haberman (2003) proposed the age-period-cohort Lee–Carter method; Hyndman and Ullah (2007) proposed a functional data model that utilizes nonparametric smoothing and high-order principal components; Girosi and King (2008) and Wiśniowski et al. (2015) considered Bayesian techniques for Lee–Carter model estimation and forecasting; and Li et al. (2013) extended the Lee–Carter method to model the rotation of age patterns for long-term projections.

These works mainly focused on forecasting age-specific mortality for a single population, or several populations individually. However, joint modeling mortality for two or more populations simultaneously is paramount, as it allows one to model the correlations among two or more populations, distinguish between long-term and short-term effects in the mortality evolution, and explore the additional information contained in the experience of other populations to further improve point and interval forecast accuracy. These populations can be grouped by sex, state, ethnic group, socioeconomic status, and other attributes (e.g., Li and Lee 2005; Alkema et al. 2011; Raftery et al. 2012, 2013; Li 2013; Raftery et al. 2014; Ševčíková et al. 2015).

As an extension of Li and Lee (2005), we consider a robust multilevel functional data model described in Sect. 2 by extending the work of Di et al. (2009), Crainiceanu

H.L. Shang (✉)
Research School of Finance, Actuarial Studies and Statistics,
Australian National University, Canberra, Australia
e-mail: hanlin.shang@anu.edu.au

C. Agostinelli et al. (eds.), *Recent Advances in Robust Statistics: Theory and Applications*, DOI 10.1007/978-81-322-3643-6_9

et al. (2009), Crainiceanu and Goldsmith (2010), Greven et al. (2010) and Shang (2016). The objective of the multilevel functional data method is to model multiple sets of functions that may be correlated among groups. In this paper, we propose a robust version of this technique to forecast age-specific mortality and life expectancy for a group of populations. We found that the robust multilevel functional data model captures the correlation among populations, models the forecast uncertainty through Bayesian paradigm, and is adequate for use within a probabilistic population modeling framework (Raftery et al. 2012). Similar to the work of Li and Lee (2005), Lee (2006), Delwarde et al. (2006), the robust multilevel functional data model captures the common trend and the population-specific residual trend. It produces forecasts that are more accurate than the ones from the standard multilevel functional data method, in the presence of outliers. Illustrated by the age- and sex-specific mortality rates for the United Kingdom, we study and compare the performance of the standard and robust multilevel functional data methods in Sect. 3. In Sect. 4, we provide some concluding remarks.

2 A Robust Multilevel Functional Data Method

We present this method in the context of forecasting female and male age-specific mortality in a country, although the method can easily be generalized to any number of subpopulations. Let $y_t^j(x_i)$ be the log central mortality of the jth population for year $t = 1, 2, \dots, n$ at observed ages $x_1, x_2, \dots, x_p$ where x is a continuous variable and p is the number of ages.

Since we consider forecasting age-specific mortality from a functional data analytic viewpoint, each function should be smooth and continuous. A nonparametric smoothing technique is thus implemented to construct a time series of functions $\left\{f_1^j(x), f_2^j(x), \dots, f_n^j(x)\right\}$. That is

$$y_t^j(x_i) = f_t^j(x_i) + \delta_t^j(x_i)\varepsilon_{t,i}^j, \tag{1}$$

where x_i represents the center of each age or age group for $i = 1, 2, \dots, p$, $\varepsilon_{t,i}^j$ is an independent and identically distributed (iid) standard normal random variable, $\delta_t^j(x_i)$ captures different variances for different ages. Together, $\delta_t^j(x_i)\varepsilon_{t,i}^j$ represents the smoothing error (also known as measurement error).

Let $m_t^j(x_i) = \exp\{y_t^j(x_i)\}$ be the observed central mortality for ages x_i at year t and let $N_t(x_i)$ be the total mid-year population of age x_i in year t. The observed mortality rate approximately follows a binomial distribution with estimated variance

$$\text{Var}\left[m_t^j(x_i)\right] \approx \frac{m_t^j(x_i) \times \left(1 - m_t^j(x_i)\right)}{N_t(x_i)}.$$

Via Taylor's series expansion, the estimated variance associated with the log mortality rate is given by

$$\left(\widehat{\delta_t^j}\right)^2(x_i) = \text{Var}\left[\ln\left(m_t^j(x_i)\right)\right] \approx \frac{1 - m_t^j(x_i)}{m_t^j(x_i) \times N_t(x_i)}.$$

Since $m_t^j(x_i)$ is often quite small, $\left(\delta_t^j\right)^2(x_i)$ can also be approximated by a Poisson distribution with estimated variance

$$\left(\widehat{\delta_t^j}\right)^2(x_i) \approx \frac{1}{m_t^j(x_i) \times N_t(x_i)}.$$

Let the weights be the inverse variances $w_t(x_i) = 1/\left(\widehat{\delta_t^j}\right)^2(x_i)$, the log mortality rates are smoothed using weighted penalized regression spline with a partial monotonic constraint for ages above 65 (Hyndman and Ullah 2007). The penalized regression spline can be written as

$$\widehat{f}_t(x_i) = \arg\min_{\theta_t(x_i)} \sum_{i=1}^{M} w_t(x_i)\left|y_t(x_i) - \theta_t(x_i)\right| + \alpha \sum_{i=1}^{M-1} \left|\theta_t^{'}(x_{i+1}) - \theta_t^{'}(x_i)\right|,$$

where i represents different ages (grid points) in a total of M grid points, α represents a smoothing parameter, and $'$ symbolizes the first derivative of a function. While the L_1 loss function and the L_1 roughness penalty are employed to obtain robust estimates, the monotonic increasing constraint helps to reduce the noise from estimation of high ages (see also He and Ng 1999).

In the multilevel functional data method, we first apply (1) to smooth different sets of functions from different populations that may be correlated. In the case of two populations, the essence is to decompose curves among two populations into an average of total mortality, denoted by $\mu(x)$, a sex-specific deviation from the averaged total mortality, denoted by $\eta^j(x)$, a common trend that is shared by all populations, denoted by $R_t(x)$, a sex-specific trend that is specific to jth population, denoted by $U_t^j(x)$, and model error $e_t^j(x)$ with finite variance $(\sigma^2)^j$. The common trend is obtained by projecting a functional time series onto the eigenvectors of covariance operators of the aggregate and centered stochastic process. The sex-specific trend is then obtained by projecting the residual functions from the first eigen decomposition, onto the eigenvectors of covariance operators of the sex-specific and centered stochastic process. To express our idea mathematically, the smoothed log mortality rates at year t can be written as follows:

$$f_t^j(x) = \mu(x) + \eta^j(x) + R_t(x) + U_t^j(x) + e_t^j(x), \qquad x \in \mathscr{I}, \tag{2}$$

for each t, where $\mathscr{I}$ represents a function support.

Since $R(x)$ and $U^j(x)$ are unknown in practice, they can be approximated by a set of realizations $\boldsymbol{R}(x) = \{R_1(x), R_2(x), \ldots, R_n(x)\}$ and $\boldsymbol{U}^j(x) = \left\{U_1^j(x), U_2^j(x), \ldots, U_n^j(x)\right\}$. Thus, the sample mean function of total mortality and sex-specific mortality can be expressed as follows:

$$\widehat{\mu}(x) = \tfrac{1}{n}\textstyle\sum_{t=1}^{n} f_t^{\mathrm{T}}(x) \tag{3}$$

$$\widehat{\mu}^j(x) = \tfrac{1}{n}\textstyle\sum_{t=1}^{n} f_t^{j}(x) \tag{4}$$

$$\widehat{\eta}^j(x) = \widehat{\mu}^j(x) - \widehat{\mu}(x) \tag{5}$$

where $\{f_1^{\mathrm{T}}(x), f_2^{\mathrm{T}}(x), \ldots, f_n^{\mathrm{T}}(x)\}$ represents a set of smooth functions for the age-specific total mortality; $\widehat{\mu}(x)$ represents the simple average of smoothed total mortality; whereas $\widehat{\mu}^j(x)$ represents the simple average of smoothed male or female mortality; and $\widehat{\eta}^j(x)$ represents the difference between the mean of total mortality and the mean of sex-specific mortality.

Then, we consider a two-step algorithm by combining a robust functional principal component analysis and binary weighting. This can be described as

(1) Use a robust principal component analysis, such as RAPCA (Hubert et al. 2002) or ROBPCA (Hubert et al. 2005), to obtain initial (highly robust) values for $\left\{\widehat{\beta}_{t,k}\right\}$ and $\left\{\widehat{\phi}_k(x)\right\}$ for $t = 1, \ldots, n$ and $k = 1, \ldots, K$.
(2) Define the integrated squared error for year t as

$$v_t = \int_{x \in \mathscr{I}} \left[f_t(x) - \sum_{k=1}^{K} \widehat{\beta}_{t,k}\widehat{\phi}_k(x)\right]^2 dx.$$

It identifies those outlying years that have higher values of v_t. We then assign weights $w_t = 1$ if $v_t < s + \lambda\sqrt{s}$ and $w_t = 0$ otherwise, where s is the median of $\{v_1, v_2, \ldots, v_n\}$ and $\lambda > 0$ is a tuning parameter to control the efficiency of this robust algorithm. When $\lambda = 3$, it represents $\Phi(3/\sqrt{2}) = 98.3\,\%$ efficiency, where the number of outliers is $1.7\,\%$ of total number of observations. When $\lambda \rightarrow \infty$, there is no outlier in the data; when $\lambda \rightarrow 0$, all observations are identified as outliers. For $\lambda > 0$, this algorithm retains the optimal breakdown point of 0.5.

Having obtained a set of robust basis functions, the common and sex-specific trends can be estimated by

$$\widehat{R}_t(x) \approx \sum_{k=1}^{K} \widehat{\beta}_{t,k}\widehat{\phi}_k(x),$$
$$\widehat{U}_t^j(x) \approx \sum_{l=1}^{L} \widehat{\gamma}_{t,l}^j \widehat{\psi}_l^j(x), \tag{6}$$

where $\left\{\widehat{\boldsymbol{\beta}}_k = \left(\widehat{\beta}_{1,k}, \widehat{\beta}_{2,k}, \ldots, \widehat{\beta}_{n,k}\right); k = 1, \ldots, K\right\}$ represents the kth sample principal component scores of $R(x)$; $\boldsymbol{\Phi} = \left[\widehat{\phi}_1(x), \widehat{\phi}_2(x), \ldots, \widehat{\phi}_K(x)\right]$ are the corresponding orthogonal sample eigenfunctions in a square integrable function space. Similarly, $\{\widehat{\boldsymbol{\gamma}}_l^j = (\widehat{\gamma}_{1,l}^j, \widehat{\gamma}_{2,l}^j, \ldots, \widehat{\gamma}_{n,l}^j); l = 1, \ldots, L\}$ represents the lth sample principal component scores of $U^j(x)$, and $\boldsymbol{\Psi} = [\widehat{\psi}_1^j(x), \widehat{\psi}_2^j(x), \ldots, \widehat{\psi}_L^j(x)]$ are the corresponding orthogonal sample eigenfunctions. Since two stochastic processes $R(x)$ and $U^j(x)$ are uncorrelated, $\widehat{\boldsymbol{\beta}}_k$ are uncorrelated with $\widehat{\boldsymbol{\gamma}}_l^j$.

It is important to select optimal K and L, and three common approaches are leave-one-out or leave-more-out cross validation (Rice and Silverman 1991), Akaike information criterion (Yao et al. 2005) and explained variance (Crainiceanu and Goldsmith 2010; Chiou 2012). We use a cumulative percentage of total variation to determine K and L. The optimal numbers of K and L are determined by

$$K = \arg\min_{K:K\geq 1}\left\{\sum_{k=1}^{K}\lambda_k \Big/ \sum_{k=1}^{\infty}\lambda_k \geq P\right\},$$

$$L = \arg\min_{L:L\geq 1}\left\{\sum_{l=1}^{L}\lambda_l^j \Big/ \sum_{l=1}^{\infty}\lambda_l^j \geq P\right\}, \qquad \text{for each } j.$$

Following Crainiceanu and Goldsmith (2010), Chiou (2012), we chose $P = 0.9$.

An important parameter in the multilevel functional data method is the proportion of variability explained by aggregate data, which is the variance explained by the within-cluster variability (Di et al. 2009). A possible measure of within-cluster variability is given by

$$\frac{\sum_{k=1}^{\infty}\lambda_k}{\sum_{k=1}^{\infty}\lambda_k + \sum_{l=1}^{\infty}\lambda_l^j} = \frac{\int_{\mathscr{I}} \text{var}\{R(x)\}dx}{\int_{\mathscr{I}} \text{var}\{R(x)\}dx + \int_{\mathscr{I}} \text{var}\{U^j(x)\}dx}. \tag{7}$$

When the common factor can explain the main mode of total variability, the value of within-cluster variability is close to 1.

Substituting Eqs. (3)–(6) into Eqs. (2) and (1), we obtain

$$y_t^j(x) = \widehat{\mu}(x) + \widehat{\eta}^j(x) + \sum_{k=1}^{K}\widehat{\beta}_{t,k}\widehat{\phi}_k(x) + \sum_{l=1}^{L}\widehat{\gamma}_{t,l}^j\widehat{\psi}_l^j(x) + e_t^j(x) + \delta_t^j(x)\varepsilon_t^j,$$

where $\widehat{\beta}_{t,k} \sim \mathrm{N}(0, \widehat{\lambda}_k)$, $\widehat{\gamma}_{t,l}^j \sim \mathrm{N}(0, \widehat{\lambda}_l^j)$, $e_t^j(x) \sim \mathrm{N}(0, (\widehat{\sigma}^2)^j)$ and $\widehat{\lambda}_k$ denotes the kth eigenvalue of estimated covariance operator associated with the common trend, and $\widehat{\lambda}_l^j$ represents the lth eigenvalue of estimated covariance operator associated with the sex-specific residual trend.

Conditioning on the estimated functional principal components $\boldsymbol{\Phi}$, $\boldsymbol{\Psi}$ and continuous functions $\boldsymbol{y}^j = [y_1^j(x), y_2^j(x), \ldots, y_n^j(x)]$, the h-step-ahead point forecasts of $y_{n+h}^j(x)$ are given by:

$$\begin{aligned}
\widehat{y}^j_{n+h|n}(x) &= \mathrm{E}\left[y_{n+h}(x)|\boldsymbol{\Phi}, \boldsymbol{\Psi}, \boldsymbol{y}^j\right] \\
&= \widehat{\mu}(x) + \widehat{\eta}^j(x) + \sum_{k=1}^{K} \widehat{\beta}_{n+h|n,k}\widehat{\phi}_k(x) + \sum_{l=1}^{L} \widehat{\gamma}^j_{n+h|n,l}\widehat{\psi}^j_l(x),
\end{aligned}$$

which $\widehat{\beta}_{n+h|n,k}$ and $\widehat{\gamma}^j_{n+h|n,l}$ are forecast univariate principal component scores, obtained from a univariate time series forecasting method, such as random walk with drift (rwf), exponential smoothing (ets), and autoregressive integrated moving average (ARIMA(p, d, q)) in which its optimal orders p, d, q are determined automatically using an information criterion, such as corrected Akaike information criterion.

If $\{\widehat{\gamma}^1_{n+h|n,l} - \widehat{\gamma}^2_{n+h|n,l}; l = 1, \dots, L\}$ has a trending long-term mean, the multilevel functional data method does not produce convergent forecasts. However, if the common mean function and common trend capture the long-term effect, the multilevel functional data method produces convergent forecasts, where the forecasts of residual trends would be flat.

To measure forecast uncertainty, the interval forecasts of $y^j_{n+h}(x)$ can be obtained through a Bayesian paradigm equipped with Markov chain Monte Carlo (MCMC). Di et al. (2009) present a derivation of posterior of principal component scores, where MCMC is used to estimate all variance parameters and to draw samples from the posterior of principal component scores. The bootstrapped forecasts are given by

$$\begin{aligned}
\widehat{y}^{b,j}_{n+h|n}(x) =& \widehat{\mu}(x) + \widehat{\eta}^j(x) + \sum_{k=1}^{K} \widehat{\beta}^b_{n+h|n,k}\widehat{\phi}_k(x) + \sum_{l=1}^{L} \widehat{\gamma}^{b,j}_{n+h|n,l}\widehat{\psi}^j_l(x) + \\
& \widehat{e}^{b,j}_{n+h}(x) + \widehat{\delta}^{b,j}_{n+h}(x)\varepsilon^j_{n+h},
\end{aligned} \tag{8}$$

for $b = 1, \dots, B$. As previously studied by Di et al. (2009, supplementary materials), we first simulate $\{\widehat{\beta}^b_{1,k}, \dots, \widehat{\beta}^b_{n,k}\}$ drawn from its posterior, and then obtain $\widehat{\beta}^b_{n+h|n,k}$ using a univariate time series forecasting method for each simulated sample; similarly, we first simulate $\{\widehat{\gamma}^{b,j}_{1,l}, \dots, \widehat{\gamma}^{b,j}_{n,l}\}$ drawn from its posterior, and then obtain $\widehat{\gamma}^{b,j}_{n+h|n,l}$ for each bootstrap sample; $\widehat{e}^{b,j}_{n+h}(x)$ is drawn from $N(0, (\widehat{\sigma}^2)^{b,j})$, where $(\widehat{\sigma}^2)^{b,j}$ is estimated at each iteration of MCMC. Since we pre-smooth functional data, we must add the smoothing error $\widehat{\delta}^{b,j}_{n+h}(x)$ as another source of randomness and ε^j_{n+h} is drawn from $N(0, 1)$ and $B = 1000$ represents the number of MCMC draws. The prediction interval is constructed from the percentiles of the bootstrapped mortality forecasts. The interval forecasts of life expectancy are obtained from the forecast age-specific mortality using the life table method (Preston et al. 2001).

3 Application to the UK's Age- and Sex-Specific Mortality

Age- and sex-specific mortality rates for the United Kingdom between 1922 and 2009 are available from the Human Mortality Database (2015). For each sex in a given calendar year, the mortality rates obtained by the ratio between "number of deaths" and "exposure to risk," are organized in a matrix by age and calendar year. By analyzing the changes in mortality as a function of age x and year t, it can be seen that age-specific mortality rates have shown a gradual decline over years. In Fig. 1a, b, we present functional time series plots of female and male log mortality rates. Using a weighted penalized regression spline, the smoothed female and male log mortality rates are obtained in Fig. 1c, d.

In the top panel of Fig. 2, we display the estimated common mean function $\widehat{\mu}(x)$, first estimated common functional principal component $\widehat{\phi}_1(x)$ and corresponding scores $\{\widehat{\beta}_{1,1}, \widehat{\beta}_{2,1}, \dots, \widehat{\beta}_{n,1}\}$ along with their 30-years-ahead forecasts. The first com-

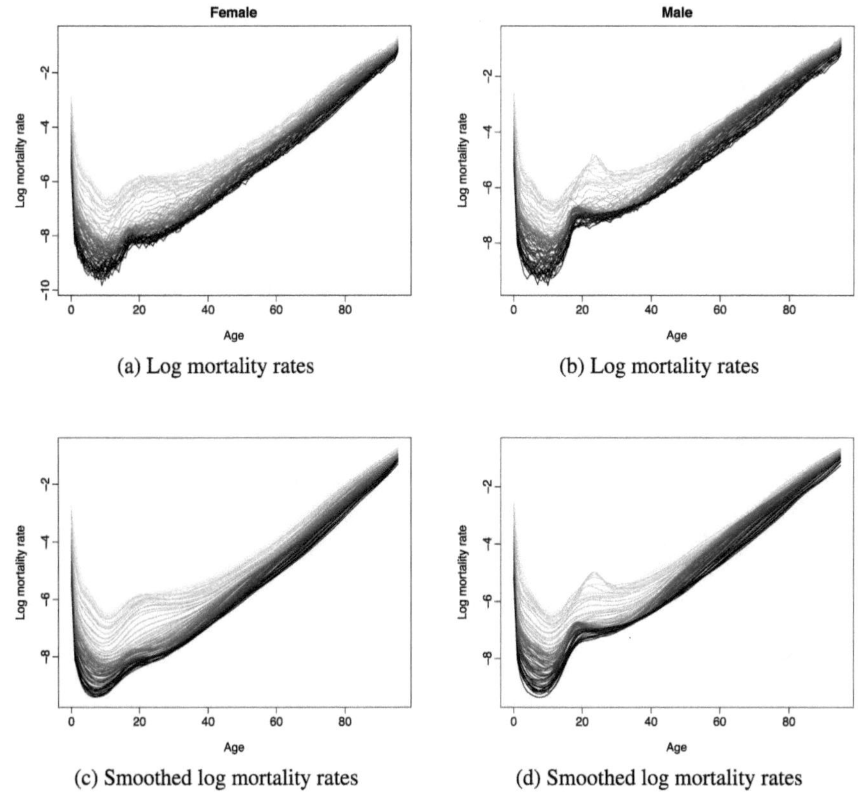

(a) Log mortality rates
(b) Log mortality rates
(c) Smoothed log mortality rates
(d) Smoothed log mortality rates

Fig. 1 Observed and smoothed age-specific female and male log mortality rates for the United Kingdom. Data from the distant past are shown in *light gray*, and the most recent data are shown in *dark gray*

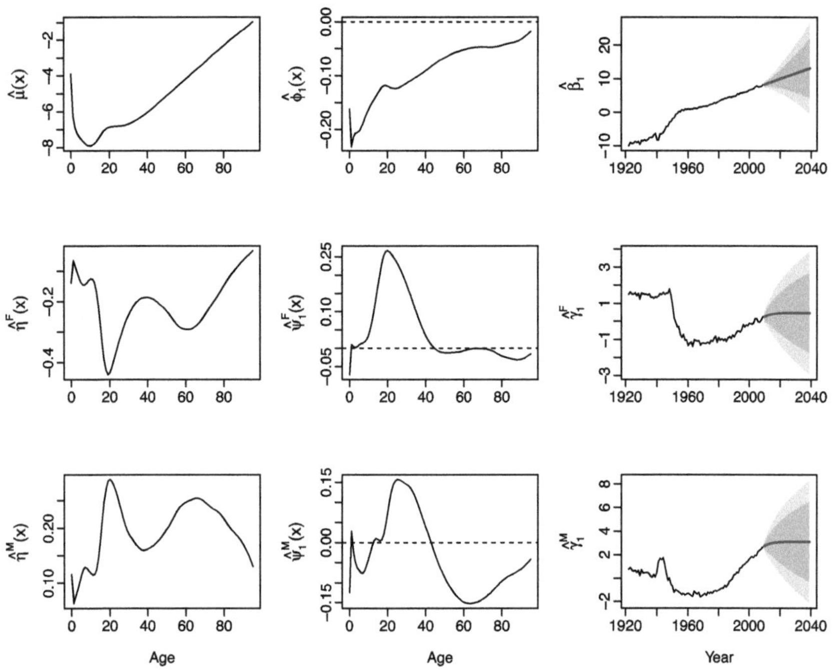

Fig. 2 Estimated common mean function, first common functional principal component, and associated scores for the UK total mortality (*top*); estimated mean function deviation for females, first functional principal component, and associated scores for the UK female mortality (*middle*); estimated mean function deviation for males, first functional principal component, and associated scores for the UK male mortality (*bottom*). The *dark* and *light gray* regions show the 80 and 95 % prediction intervals, respectively

mon functional principal component captures more than 98 % of the total variation in the age-specific total mortality. In the middle panel of Fig. 2, we show the estimated mean function deviance of females from the overall mean function $\widehat{\eta}^{\mathrm{F}}(x)$, first functional principal component for females $\widehat{\psi}_1^{\mathrm{F}}(x)$ and corresponding scores $\{\widehat{\gamma}_{1,1}^{\mathrm{F}}, \widehat{\gamma}_{2,1}^{\mathrm{F}}, \ldots, \widehat{\gamma}_{n,1}^{\mathrm{F}}\}$ with 30-years-ahead forecasts. In the bottom panel of Fig. 2, we display the estimated mean function deviance of males from the overall mean function $\widehat{\eta}^{\mathrm{M}}(x)$, first functional principal component for males $\widehat{\psi}_1^{\mathrm{M}}(x)$ and corresponding scores $\{\widehat{\gamma}_{1,1}^{\mathrm{M}}, \widehat{\gamma}_{2,1}^{\mathrm{M}}, \ldots, \widehat{\gamma}_{n,1}^{\mathrm{M}}\}$ with 30-years-ahead forecasts. In this data set, the first three functional principal components explain at least 90 % of the remaining 10 % total variations for both females and males. Here, we display only the first functional principal component, which captures more than 64 and 50 % of the remaining 10 % total variations for both females and males, respectively. Based on (7), the proportion of variability explained by the total mortality is 94 % for females and 95 % for males.

3.1 *Forecast Accuracy Evaluation*

3.1.1 Evaluation of Point Forecast Accuracy

We split our age- and sex-specific data into a training sample (including data from years 1 to $(n-30)$) and a testing sample (including data from years $(n-29)$ to n), where n represents the total number of years in the data. Following the early work by Hyndman and Booth (2008), we implement an expanding window approach as it allows us to assess the forecast accuracy among methods for different forecast horizons. With the initial training sample, we produce one-to 30-year-ahead forecasts, and determine the forecast errors by comparing the forecasts with actual out-of-sample data. As the training sample increases by one year, we produce one-to 29-year-ahead forecasts and calculate the forecast errors. This process continues until the training sample covers all available data.

To measure the point forecast accuracy, we utilize the root mean squared forecast error (RMSFE), root maximum squared forecast error (Max RSFE), mean absolute forecast error (MAFE), maximum absolute forecast error (Max AFE), and mean forecast error (MFE). They are defined as

$$\text{RMSFE}(h) = \sqrt{\frac{1}{(31-h)\times p}\sum_{k=n-30+h}^{n}\sum_{i=1}^{p}[m_k(x_i)-\widehat{m}_k(x_i)]^2},$$

$$\text{Max RSFE}(h) = \sqrt{\max_{k,i}[m_k(x_i)-\widehat{m}_k(x_i)]^2},$$

$$\text{MAFE}(h) = \frac{1}{(31-h)\times p}\sum_{k=n-30+h}^{n}\sum_{i=1}^{p}|m_k(x_i)-\widehat{m}_k(x_i)|,$$

$$\text{Max AFE}(h) = \max_{k,i}|m_k(x_i)-\widehat{m}_k(x_i)|,$$

$$\text{MFE}(h) = \frac{1}{(31-h)\times p}\sum_{k=n-30+h}^{n}\sum_{i=1}^{p}[m_k(x_i)-\widehat{m}_k(x_i)],$$

for $k=n-30+h,\dots,n$ and $h=1,\dots,30$, where $m_k(x_i)$ represents mortality rate at year k in the forecasting period for age x_i, and $\widehat{m}_k(x_i)$ represents the point forecast.

3.1.2 Evaluation of Interval Forecast Accuracy

To assess interval forecast accuracy, we use the interval score of Gneiting and Raftery (2007) (see also Gneiting and Katzfuss 2014). For each year in the forecasting period, one-year-ahead to 30-year-ahead prediction intervals were calculated at the $(1-\alpha)\times 100\,\%$ prediction interval, with lower and upper bounds that are predictive quantiles at $\alpha/2$ and $1-\alpha/2$, denoted by x_l and x_u. As defined by Gneiting and Raftery (2007), a scoring rule for the interval forecast at age x_i is

$$S_\alpha(x_l, x_u; x_i) = (x_u - x_l) + \frac{2}{\alpha}(x_l - x_i)I\{x_i < x_l\} + \frac{2}{\alpha}(x_i - x_u)I\{x_i > x_u\}, \quad (9)$$

where $I\{\cdot\}$ represents the binary indicator function, and α denotes the level of significance, customarily $\alpha = 0.2$. A forecaster is rewarded for narrow prediction intervals, but incurs a penalty, the size of which depends on α, if the observation misses the interval. The smallest interval score is the one that achieves the best trade off between empirical coverage probability and halfwidth of prediction interval.

For different ages and years in the forecasting period, the maximum and mean interval scores for each horizon are defined by

$$\max[S_\alpha(h)] = \max_{k,i} S_{\alpha,k}(x_l, x_u; x_i),$$

$$\bar{S}_\alpha(h) = \frac{1}{(31-h) \times p} \sum_{k=n-30+h}^{n} \sum_{i=1}^{p} S_{\alpha,k}(x_l, x_u; x_i),$$

where p represents the total number of ages or age groups in the evaluation data set. The best forecasting method is considered to be the one that produces the smallest maximum or mean interval score.

3.2 Comparison of Point Forecast Accuracy

We compare the point forecast accuracy between the standard and robust multilevel functional data methods. As with the robust multilevel functional data method, it is necessary to specify a tuning parameter λ. When $\lambda \to \infty$, it corresponds to the standard multilevel functional data method, where no outlier can be detected. When $\lambda \to 0$, it considers all observations as outliers. Here, we consider four different values for $\lambda = 1.81, 2.33, 3, 3.29$, which reflects 90, 95, 98.3, and 99 % of efficiency. For this data set, we found that the robust multilevel functional data method outperforms the standard multilevel functional data method. The optimal forecast accuracy is achieved when $\lambda = 1.81$, regardless which univariate time series forecasting method (rwf, ARIMA, ets) is used. Among the three univariate time series forecasting methods, the random walk with drift generally performs the best with the smallest forecast errors for female and male mortality rates and male life expectancy, whereas the ARIMA forecasting method produces the smallest forecast errors for female life expectancy (Table 1).

Table 1 Point forecast accuracy of age-specific mortality and life expectancy for females and males by different univariate time series forecasting methods, as measured by the Max AFE, Max RSFE, MAFE, RMSFE, and MFE. For mortality, the forecast errors were multiplied by 100 in order to keep two decimal places. The minimal forecast errors are highlighted in bold for females and males

Error	Sex	Mortality (×100)					Life expectancy				
		1.81	2.33	3	3.29	∞	1.81	2.33	3	3.29	∞
MaxAFE	F	6.11	6.18	6.25	6.29	7.18	2.23	2.27	2.33	2.34	3.24
(rwf)	M	**7.31**	7.36	7.41	7.45	8.68	**2.27**	2.31	2.38	2.41	3.15
	F	**6.09**	6.15	6.22	6.19	7.13	**2.09**	2.16	2.16	2.16	3.04
(ARIMA)	M	7.90	7.93	8.06	8.10	8.55	2.88	2.88	2.97	2.98	3.23
	F	6.50	6.58	6.62	6.64	7.60	2.74	2.79	2.80	2.80	3.65
(ets)	M	7.99	8.01	8.09	8.07	9.10	3.36	3.40	3.60	3.56	4.00
Max RSFE	F	**0.38**	0.39	0.40	0.40	0.52	5.56	5.76	6.03	6.11	10.61
(rwf)	M	**0.57**	0.58	0.59	0.59	0.79	**6.26**	6.46	6.84	6.99	10.95
	F	**0.38**	0.39	0.40	0.39	0.51	**4.83**	5.13	5.17	5.16	9.29
(ARIMA)	M	0.66	0.67	0.69	0.70	0.78	10.00	9.93	10.45	10.54	11.57
	F	0.43	0.44	0.45	0.45	0.58	8.56	8.83	8.87	8.89	13.64
(ets)	M	0.69	0.69	0.71	0.70	0.88	13.72	14.02	15.55	15.23	17.92
MAFE	F	**0.42**	0.43	0.46	0.48	0.68	1.59	1.63	1.73	1.78	2.53
(rwf)	M	**0.61**	0.62	0.66	0.67	0.82	**1.76**	1.80	1.90	1.94	2.47
	F	0.45	0.46	0.48	0.48	0.64	**1.48**	1.54	1.60	1.64	2.22
(ARIMA)	M	0.73	0.74	0.77	0.79	0.84	2.16	2.16	2.25	2.29	2.56
	F	0.50	0.51	0.54	0.55	0.75	2.02	2.06	2.15	2.18	2.88
(ets)	M	0.78	0.79	0.83	0.84	0.95	2.56	2.59	2.68	2.71	3.11
RMSFE	F	**0.92**	0.95	1.01	1.03	1.39	1.66	1.71	1.80	1.85	2.56
(rwf)	M	**1.23**	1.25	1.32	1.34	1.58	**1.81**	1.85	1.95	1.99	2.52
	F	0.99	1.02	1.04	1.05	1.34	**1.54**	1.61	1.66	1.70	2.27

(continued)

Table 1 (continued)

Error	Sex	Mortality (×100)					Life expectancy				
		1.81	2.33	3	3.29	∞	1.81	2.33	3	3.29	∞
(ARIMA)	M	1.44	1.44	1.51	1.53	1.61	2.25	2.24	2.33	2.37	2.61
	F	1.06	1.09	1.14	1.16	1.53	2.09	2.13	2.20	2.24	2.91
(ets)	M	1.50	1.53	1.59	1.61	1.81	2.63	2.67	2.75	2.78	3.16
MFE	F	**−0.33**	−0.35	−0.39	−0.41	−0.67	1.58	1.62	1.72	1.77	2.53
(rwf)	M	**−0.36**	−0.38	−0.45	−0.47	−0.78	**1.71**	1.75	1.86	1.91	2.47
	F	−0.37	−0.39	−0.41	−0.42	−0.63	**1.47**	1.53	1.59	1.63	2.22
(ARIMA)	M	−0.52	−0.52	−0.58	−0.61	−0.80	2.09	2.10	2.20	2.24	2.56
	F	−0.44	−0.45	−0.48	−0.50	−0.74	2.02	2.05	2.14	2.18	2.88
(ets)	M	−0.60	−0.62	−0.67	−0.69	−0.92	2.51	2.54	2.65	2.68	3.11

Table 2 Interval forecast accuracy of mortality and life expectancy for females and males by different univariate time series forecasting methods, as measured by maximum interval score and mean interval score. For mortality, the interval scores were multiplied by 100 in order to keep two decimal places. The minimal interval scores are highlighted in bold for females and males

Error	Sex	Mortality (×100)					Life expectancy				
		1.81	2.33	3	3.29	∞	1.81	2.33	3	3.29	∞
Max interval score	F	26.17	25.72	27.64	27.93	33.40	11.84	12.23	13.09	13.09	18.26
(rwf)	M	**25.12**	26.72	28.73	29.77	48.94	9.86	10.16	11.14	11.18	16.66
	F	17.20	16.13	17.30	17.34	18.89	6.13	**5.92**	7.74	7.60	6.52
(ARIMA)	M	33.39	33.31	33.39	33.75	38.01	15.37	15.04	16.86	16.71	12.04
	F	16.06	15.57	15.50	**15.07**	18.24	6.72	6.79	6.76	6.42	7.25
(ets)	M	26.39	27.40	26.73	26.40	42.81	**8.35**	9.25	9.25	9.84	12.96
Mean interval score	F	2.27	2.36	2.57	2.69	3.42	8.04	8.47	9.25	9.66	13.50
(rwf)	M	3.11	3.23	3.49	3.62	4.38	8.18	8.57	9.21	9.56	12.35
	F	1.49	**1.48**	1.57	1.62	1.52	4.75	**4.67**	5.09	5.36	4.97
(ARIMA)	M	3.68	3.50	3.74	3.62	3.11	11.52	10.76	11.88	11.13	9.40
	F	1.56	1.58	1.56	1.51	1.48	5.95	5.84	5.83	5.50	5.35
(ets)	M	**2.91**	3.01	2.98	3.08	3.30	**7.38**	7.88	7.82	8.08	8.54

3.3 Comparison of Interval Forecast Accuracy

The prediction intervals for age-specific mortality are obtained from (8), whereas the prediction intervals for life expectancy are obtained from the percentiles of simulated life expectancies obtained from simulated forecast mortality rates as described by Hyndman and Booth (2008). Based on the mean interval scores in Table 2, we found the robust multilevel functional data method outperforms the standard multilevel functional data method. The ARIMA forecasting method gives the smallest interval scores for females when $\lambda = 2.33$, whereas the exponential smoothing method performs the best for males when $\lambda = 1.81$.

4 Conclusion

In this paper, we put forward a robust multilevel functional data method to forecast age-specific mortality and life expectancy at birth for a group of populations. This method inherits the smoothness property a functional time series possesses, thus missing data can be naturally dealt with. In addition, this method is a robust approach that can handle the presence of outliers.

As demonstrated by the empirical studies consisting of two subpopulations in the UK, we found that the robust multilevel functional data method produces more accurate forecasts than the standard multilevel functional data method in the presence of outlying years largely due to World Wars and Spanish flu pandemic in the UK. Based on the averaged forecast errors, the robust multilevel functional data method with $\lambda = 1.81$ gives the most accurate point forecasts among all we considered. Furthermore, we consider three univariate time series forecasting methods and compare their point and interval forecast accuracy. Among the three univariate time series forecasting methods, the random walk with drift generally performs the best for female and male mortality rates and male life expectancy, whereas the ARIMA forecasting method produces the smallest point forecast errors for female life expectancy. Based on the mean interval scores, the ARIMA forecasting method gives the smallest interval scores for females when $\lambda = 2.33$, whereas the exponential smoothing method performs the best for males when $\lambda = 1.81$. It is a straightforward extension to average forecasts obtained from all three univariate time series forecasting methods in hope to improve forecast accuracy. Although $\lambda = 1.81$ works well in the data set considered, the optimal selection of λ remains as a challenge and an open problem for future research.

Other research topics are that although the proposed methods are demonstrated using the UK data, the methodology can easily be extended to mortality data from other countries. Furthermore, the multilevel functional data model captures correlation between a group of populations based on sex, but the methodology can also be extended to some other characteristics, such as state or ethnic group. It would also be interesting to investigate the performance of this robust multilevel functional data method for various lengths of functional time series.

Acknowledgments The author is grateful for the invitation by Professor Graciela Boente to participate the ICORS2015 conference. The author thanks comments and suggestions received from the participants of the ICORS2015 conference, and the participants of the Bayesian methods for population estimation workshop held at the Australian Bureau of Statistics in May, 2015.

References

Alkema L, Raftery AE, Gerland P, Clark SJ, Pelletier F, Buettner T, Heilig GK (2011) Probabilistic projections of the total fertility rate for all countries. Demography 48(3):815–839

Booth H (2006) Demographic forecasting: 1980–2005 in review. Int J Forecast 22(3):547–581

Booth H, Tickle L (2008) Mortality modelling and forecasting: a review of methods. Ann Actuarial Sci 3(1–2):3–43

Chiou JM (2012) Dynamical functional prediction and classification, with application to traffic flow prediction. Ann Appl Statis 6(4):1588–1614

Crainiceanu CM, Goldsmith JA (2010) Bayesian functional data analysis using WinBUGS. J Statis Softw 32(11)

Crainiceanu CM, Staicu AM, Di CZ (2009) Generalized multilevel functional regression. J Am Statis Assoc 104(488):1550–1561

Delwarde A, Denuit M, Guillén M, Vidiella-i-Anguera A (2006) Application of the Poisson log-bilinear projection model to the G5 mortality experience. Belgian Actuarial Bull 6(1):54–68

Di CZ, Crainiceanu CM, Caffo BS, Punjabi NM (2009) Multilevel functional principal component analysis. Ann Appl Statis 3(1):458–488

Girosi F, King G (2008) Demographic forecasting. Princeton University Press, Princeton

Gneiting T, Katzfuss M (2014) Probabilistic forecasting. Ann Rev Statis Appl 1:125–151

Gneiting T, Raftery AE (2007) Strictly proper scoring rules, prediction and estimation. J Am Statis Assoc 102(477):359–378

Greven S, Crainiceanu C, Caffo B, Reich D (2010) Longitudinal functional principal component analysis. Electron J Statis 4:1022–1054

He X, Ng P (1999) COBS: qualitatively constrained smoothing via linear programming. Comput Statis 14:315–337

Hubert M, Rousseeuw P, Verboven S (2002) A fast method for robust principal components with applications to chemometrics. Chemom Intell Lab Syst 60(1–2):101–111

Hubert M, Rousseeuw P, Branden K (2005) ROBPCA: a new approach to robust principal component analysis. Technometrics 47(1):64–79

Human Mortality Database (2015) University of California, Berkeley (USA), and Max Planck Institute for Demographic Research (Germany). Accessed 8 March 2013. http://www.mortality.org

Hyndman RJ, Booth H (2008) Stochastic population forecasts using functional data models for mortality, fertility and migration. Int J Forecast 24(3):323–342

Hyndman RJ, Ullah MS (2007) Robust forecasting of mortality and fertility rates: a functional data approach. Comput Statis Data Anal 51(10):4942–4956

Lee RD (2006) Mortality forecasts and linear life expectancy trends. In: Bengtsson T (ed) Perspectives on mortality forecasting, vol III. The linear rise in life expectancy: History and prospects, no. 3 in Social Insurance Studies, Swedish National Social Insurance Board, Stockholm, pp 19–39

Lee RD, Carter LR (1992) Modeling and forecasting U.S. mortality. J Am Statis Assoc 87(419):659–671

Li J (2013) A Poisson common factor model for projecting mortality and life expectancy jointly for females and males. Popul Stud 67(1):111–126

Li N, Lee R (2005) Coherent mortality forecasts for a group of population: an extension of the Lee-Carter method. Demography 42(3):575–594

Li N, Lee R, Gerland P (2013) Extending the Lee-Carter method to model the rotation of age patterns of mortality decline for long-term projections. Demography 50(6):2037–2051

Preston SH, Heuveline P, Guillot M (2001) Demography: measuring and modelling population process. Blackwell, Oxford

Raftery AE, Li N, Ševčíková H, Gerland P, Heilig GK (2012) Bayesian probabilistic population projection for all countries. Proc Natl Acad Sci USA 109(35):13,915–13,921

Raftery AE, Chunn JL, Gerland P, Ševčíková H (2013) Bayesian probabilistic projections of life expectancy for all countries. Demography 50(3):777–801

Raftery AE, Lalic N, Gerland P (2014) Joint probabilistic projection of female and male life expectancy. Demographic Res 30:795–822

Renshaw AE, Haberman S (2003) Lee-Carter mortality forecasting with age-specific enhancement. Insur.: Math Econ 33(2):255–272

Rice J, Silverman B (1991) Estimating the mean and covariance structure nonparametrically when the data are curves. J R Statis Soc Ser B 53(1):233–243

Ševčiková H, Li N, Kantorová V, Gerland P, Raftery AE (2015) Age-specific mortality and fertility rates for probabilistic population projections. Working Paper, University of Washington. http://arxiv.org/pdf/1503.05215v1.pdf

Shang HL (2016) Mortality and life expectancy forecasting for a group of populations in developed countries: a multilevel functional data method. Ann Appl Stat, (in press)

Shang HL, Booth H, Hyndman RJ (2011) Point and interval forecasts of mortality rates and life expectancy: a comparison of ten principal component methods. Demographic Res 25(5):173–214

Tickle L, Booth H (2014) The longevity prospects of Australian seniors: an evaluation of forecast method and outcome. Asia-Pacific J Risk Insur 8(2):259–292

Wiśniowski A, Smith PWF, Bijak J, Raymer J, Forster JJ (2015) Bayesian population forecasting: extending the Lee-Carter method. Demography 52(3):1035–1059

Yao F, Müller HG, Wang J (2005) Functional data analysis for sparse longitudinal data. J Am Statis Assoc 100(470):577–590

Asymptotically Stable Tests with Application to Robust Detection

Georgy Shevlyakov

1 Introduction

Let $X_1, \ldots, X_n$ be i.i.d. observations from a distribution F with density $f(x, \theta)$, where θ is an unknown scalar parameter. Consider the problem of testing the null hypothesis

$$H_0 : \theta = \theta_0$$

against the one-sided alternative

$$H_1 : \theta > \theta_0$$

with the decision rule based on the comparison of a test statistic $T_n(X_1, \ldots, X_n)$ with a critical value $\gamma_{1-\alpha}$: reject H_0 if and only if $T_n(X_1, \ldots, X_n) > \gamma_{1-\alpha}$.

To define the asymptotic power of this test, we consider a decreasing sequence of alternatives $\theta_n = \theta_0 + A/\sqrt{n}$ where $A > 0$ (Noether 1967; Hossjer and Mettiji 1993). In this paper, we use M-estimators of the parameter θ defined by the estimating equation (Huber 1964)

$$\sum \psi(X_i, T_n) = 0 \tag{1}$$

as a test statistic in the decision rule

$$n^{1/2}\, T_n(X_1, \ldots, X_n) > \lambda_{1-\alpha}. \tag{2}$$

Here $\psi(x, \theta)$ is a score function; $\lambda_{1-\alpha} = \Phi^{-1}(1-\alpha)\, V^{1/2}(\psi, f)$ is a threshold providing the required type I error rate α (Hampel et al. 1986); under regularity

G. Shevlyakov (✉)
Department of Applied Mathematics, Peter the Great St. Petersburg Polytechnic University, St. Petersburg, Russia
e-mail: Georgy.Shevlyakov@phmf.spbstu.ru

C. Agostinelli et al. (eds.), *Recent Advances in Robust Statistics: Theory and Applications*, DOI 10.1007/978-81-322-3643-6_10

conditions (Huber 1964; Michel and Pfanzagle 1971), the asymptotic variance of $n^{1/2}T_n$ is of the following form:

$$V(\psi,f) = \left. \frac{\int \psi(x;\theta)^2 f(x;\theta)\,dx}{\left(\int \partial\psi(x;\theta)/\partial\theta\, f(x;\theta)\,dx\right)^2} \right|_{\theta=\theta_0}; \tag{3}$$

$\Phi(x) = (2\pi)^{-1/2}\int_{-\infty}^{x} e^{-t^2/2}\,dt$ is the standard Gaussian distribution function. In the case of a location parameter θ, the asymptotic variance (3) was first derived in (Huber 1964); later this result was generalized in (Michel and Pfanzagle 1971) for an arbitrary parameter θ.

The power of the Neyman–Pearson test (2) is given by (Hampel et al. 1986, p. 194)

$$\beta(\psi,f) = 1 - \Phi\left(\Phi^{-1}(1-\alpha) - AV^{-1/2}(\psi,f)\right). \tag{4}$$

In (Shurygin 1994a, b), the so-called *variational optimization approach* to the design of stable estimators is proposed. In what follows, we comment on the connection of this approach with the basic concepts of the calculus of variations, as well as with the classical approach to robustness based on the change-of-variance function (Hampel et al. 1986). These issues are also discussed in (Shevlyakov et al. 2008).

The variational optimization approach generally leads to redescending M-estimators (1), which were originally proposed in the well-known Princeton experiment (Andrews et al. 1972) from heuristic considerations and only later theoretically justified within Hampel's approach to robust estimation: on the whole, commonly used redescenders outperform the bounded M-estimators of Huber's type (for example, see (Hampel et al. 1986, p. 167, Table 3)). The variational optimization approach gives another way of the justification of redescending M-estimators being mathematically based on the solutions of non-standard problems of the calculus of variations. Its statistical interpretation is also different from the standard robustness philosophy—in the sequel, we will discuss these issues. Although this approach was originally proposed about twenty years ago, it still remains unknown to the majority of the statistical community, and thus, an attempt to expose it with the examples of its fruitful applications, perhaps, may be called a main contribution of this paper.

An outline of the remainder of the paper is as follows. In Sect. 2, an extensive review of the variational optimization approach is given. In Sect. 3, we briefly introduce the error sensitivity and stability of a test, which are defined so that the tools developed for stable estimation can be directly applied to the design of stable tests. In Sect. 4, the proposed stable tests for location based on redescending M-estimators are used for robust detection of a weak signal: they outperform both Huber's conventional and Hampel's robust tests under heavy-tailed distributions. In Sect. 5, the connections of the proposed approach with the conventional robust methods are discussed, and some conclusions are drawn.

2 Stable Estimation: A Review

The material presented in this section mostly originates from (Shurygin 1994a, b; Shevlyakov et al. 2008; Shurygin 2009).

2.1 *The Asymptotic Variance of M-Estimators*

As the initial point of all further theoretical constructions is the asymptotic variance of M-estimators, we begin with the conditions under which Eq. (3) holds (Bentkus et al. 1995).

Let $\chi = \{x : f(x, \theta) > 0\}$ denote the support of density $f(x, \theta)$ (independent of θ). For an open subset T of $\mathbb{R}$, let $\psi : \chi \times T \to \mathbb{R}$ be a jointly measurable function. Denote the partial derivative of the score function $\psi(x, t)$ with respect to t as $\dot{\psi} = \dot{\psi}(x, t) = (\partial/\partial t)\psi(x, t)$.

Theorem 2.1 (Shurygin 2009) *Let the following assumptions hold.*

1. *In the neighborhood $T \ni \theta$, the function $\psi(x, t)$ is continuous and continuously differentiable with respect to both variables almost everywhere.*
2. *For $t \in T$, $|E\dot{\psi}(x, t)|$ is positive and bounded.*
3. *The function $E\psi(x, t)^2$ is continuous at $t = \theta$.*
4. *In the neighborhood $T \ni \theta$, $E\psi(x, t)^2 > 0$.*
5. *The M-estimator T_n is consistent.*

Then

$$E\psi(x, \theta) = \int_{\chi} \psi(x, \theta) f(x, \theta)\, dx = 0, \tag{5}$$

the distribution of $\sqrt{n}(T_n - \theta)$ converges to the normal distribution $N(0, V)$ with the variance

$$V = V(\psi, f) = \frac{E\psi(x, \theta)^2}{\left(\dfrac{d}{dt} E\psi(x, t)\Big|_{t=\theta}\right)^2}; \tag{6}$$

for the continuously differentiable $\psi(x, t)$, this variance is equal to

$$V = V(\psi, f) = \frac{E\psi(x, \theta)^2}{\left(E\dot{\psi}(x, \theta)\right)^2}. \tag{7}$$

2.2 *Methods of Functional Optimization*

In what follows, we aim at optimizing the indicators of the accuracy and stability of estimation. In particular, we will mostly deal with the asymptotic variance of

M-estimators that depends on the functions $\psi(x, \theta)$ and $f(x, \theta)$ of the two variables: $x \in \chi$ and $\theta \in \Theta$. In this case, optimization is restricted to searching the minima of integrals of the following form:

$$J(\psi) = \int_\chi K(x, \theta, \psi, \dot{\psi})\, dx$$

where

$$K(x, \theta, \psi, \dot{\psi}) = Q(x, \theta, \psi) + \lambda f(x, \theta)\dot{\psi}\,,$$

$\lambda = \lambda(\theta)$, the score function $\psi(x, \theta)$ and the parameter θ are connected by the "condition of consistency" (5)

$$\int_\chi \psi(x, \theta) f(x, \theta)\, dx = 0, \tag{8}$$

and $Q(x, \theta, \psi)$ is a positively definite quadratic polynomial of ψ.

Unlike the classical case, here integration is performed by the variable x and differentiation by the variable θ. The solution of this nonstandard variational problem is lightened by condition (8). The variations, satisfying this condition, will be called *admissible*.

Lemma 2.1 (Shurygin 2009) *Let $\psi(x, \theta)$ be continuously differentiable and satisfy condition (8). Then there exists a unique, up to the factor independent of x, solution to the variational problem of minimization of the functional*

$$J(\psi) = \int_\chi K(x, \theta, \psi, \dot{\psi})\, dx$$

under the admissible variations $\delta\psi$ of a score function, and the necessary condition of minimum is given by

$$\frac{\partial Q}{\partial \psi} + \lambda_0 f(x, \theta) - \lambda_1 \dot{f}(x, \theta) = 0\,, \tag{9}$$

where the coefficients λ_0 and λ_1 do not depend on x.

Proof Taking into account the condition of consistency (8), rewrite the functional $J(\psi)$ with the corresponding Lagrange multiplier λ_0

$$J(\psi) = \int_\chi [Q + \lambda_0 \psi f + \lambda_1 \dot{\psi} f]\, dx\,.$$

Compute the variation of the functional $J(\psi)$ under the admissible increment of the score function $\delta\psi$

$$\delta J(\psi) = \int_{\mathcal{X}} \left[\frac{\partial Q}{\partial \psi} \delta\psi + \lambda_0 f \delta\psi + \lambda_1 f \delta\dot{\psi} \right] dx\,.$$

The admissible increments satisfy the equality $\int_{\mathcal{X}} f \delta\psi \, dx = 0$, which after differentiation by θ yields $\int_{\mathcal{X}} (\dot{f}\delta\psi + f\delta\dot{\psi})\, dx = 0$, so that $\int_{\mathcal{X}} f\delta\dot{\psi}\, dx = -\int_{\mathcal{X}} \dot{f}\delta\psi\, dx$. Substituting this equality into the expression for $\delta J(\psi)$, we get it in the following form at the stationary function ψ

$$\delta J(\psi) = \int_{\mathcal{X}} \left[\frac{\partial Q}{\partial \psi} + \lambda_0 f(x,\theta) - \lambda_1 \dot{f}(x,\theta) \right] \delta\psi dx = 0\,.$$

Thus, we have arrived at the standard variational problem. Apparently, the function ψ, which solves the linear equation $[\cdot] = 0$, is just the stationary function given by (9). The uniqueness of this solution directly follows from the assumed quadratic form of the functional.

2.3 Stable M-Estimators

In order to design stable estimators, we need a measure of the instability or sensitivity of estimation, which could be minimized by the appropriate choice of a score function. Similarly to the definitions of the influence function (Hampel 1974), the sensitivity curve (Tukey 1977) and the change-of-variance function (Hampel et al. 1986), a natural choice of this measure is given by some kind of a functional derivative, as just the derivative of a function is the standard indicator of its changes. Shurygin (1994a, b) proposed to use the so-called functional Lagrange derivative of the asymptotic variance $V(\psi, f)$ (7) with respect to density f under the fixed score function ψ formally defined as

$$\frac{\partial V(\psi, f)}{\partial f} = \frac{\partial}{\partial f} \frac{\int_{\mathcal{X}} \psi^2 f\, dx}{\left(\int_{\mathcal{X}} \dot{\psi} f\, dx\right)^2} = \frac{\int_{\mathcal{X}} \psi^2\, dx}{\left(\int_{\mathcal{X}} \dot{\psi} f\, dx\right)^2} - 2\frac{\int_{\mathcal{X}} \dot{\psi}\, dx \int_{\mathcal{X}} \psi^2 f\, dx}{\left(\int_{\mathcal{X}} \dot{\psi} f\, dx\right)^3}\,. \tag{10}$$

The score function ψ is defined up to the factor independent of x, that is, an M-estimator is defined by the class of score functions $\Psi = \{\psi_1(x,\theta) = c(\theta)\psi(x,\theta)\}$, where the constant $c(\theta) \neq 0$. Now, we show that there exists an element from this class, at which the second summand in (10) vanishes (Shurygin 2009).

Consider the second summand in the expression for the functional derivative (10)

$$Z = -2 \int_{\mathcal{X}} \dot{\psi}\, dx \int_{\mathcal{X}} \psi^2 f\, dx \left(\int_{\mathcal{X}} \dot{\psi} f\, dx \right)^{-3}$$

and show that it is always possible to find a factor $c = c(\theta) \neq 0$ such that $\int_X \dot{\psi}_1\, dx = 0$. Let us find the solution of this equation with respect to c

$$\int_X \dot{\psi}_1\, dx = \int_X (\dot{c}\psi + c\dot{\psi})\, dx = \dot{c}\int_X \psi\, dx + c\int_X \dot{\psi}\, dx = 0\,.$$

If $\int_X \psi\, dx = 0$, then $\dfrac{\partial}{\partial\theta}\displaystyle\int_X \psi\, dx = 0$, that is, $\int_X \dot{\psi}_1\, dx = 0$ for any value of $c(\theta)$.

If $\int_X \psi\, dx \neq 0$, then set $c(\theta) = \left(\int_X \psi\, dx\right)^{-1}$ and get that

$$\int_X \dot{\psi}_1\, dx = \int_X (\dot{c}\psi + c\dot{\psi})\, dx = \int_X \left[-\frac{\psi \int_X \dot{\psi}\, dx}{\left(\int_X \psi\, dx\right)^2} + \frac{\dot{\psi}}{\int_X \psi\, dx}\right] dx = 0\,.$$

So, $Z = 0$ also in this case.

It is convenient to take the first summand of (10), which is independent of the factor $c(\theta)$, as a measure of the sensitivity of M-estimators.

Definition 2.1 (Shurygin 1994a, b) The *variance sensitivity* of the M-estimator (1) (to possible changes of density f), corresponding to the score function $\psi(x,\theta)$, is defined as

$$VS(\psi, f) = \frac{\partial V(\psi, f)}{\partial f} = \frac{\int_X \psi(x,\theta)^2\, dx}{\left(\int_X \dot{\psi}(x,\theta) f(x,\theta)\, dx\right)^2}\,. \tag{11}$$

Definition 2.2 An M-estimator T_n and its score function $\psi(x,\theta)$ satisfying condition (8) are called *stable*, if the variance sensitivity is bounded and *unstable* otherwise.

Theorem 2.2 (Shurygin 2009) *Let Ψ_1 be a set of stable score functions $\psi(x,\theta)$. Then the variance sensitivity $VS(\psi, f)$ is positive and attains its minimum VS_{min} at the* minimum variance sensitivity *(MVS) score function*

$$\psi_{MVS}(x,\theta) = \arg\min_{\psi\in\Psi_1} VS(\psi, f) = c\left(\frac{\partial}{\partial\theta} f(x,\theta) + \beta f(x,\theta)\right),$$

where $c \neq 0$ is an arbitrary constant and β is a constant providing condition (8):

$$\beta = \beta_{MVS} = -\frac{1}{2}\frac{d}{d\theta}\log\int_X f(x,\theta)^2 dx\,.$$

Proof The assertion of this theorem directly follows from Lemma 2.1.

Now, we consider the example of stable estimation of a location parameter (a scale parameter scale is assumed known).

Corollary 2.1 *If a distribution density has the form*

$$f(x-\mu,\sigma)=\frac{1}{\sigma}h\left(\frac{x-\mu}{\sigma}\right),\quad -\infty<x,\mu<\infty,\quad \sigma>0,$$

then the MVS estimate of μ *satisfies the following equation*

$$\sum \frac{\partial}{\partial\mu}f(x_i-\mu,\sigma)\Big|_{\mu=\mu_{MVS}}=0$$

with the following minimum variance sensitivity

$$VS_{min}=VS(\psi_{MVS},f)=\left(\int_{-\infty}^{\infty}\psi_{MVS}(x-\mu)^2dx\right)^{-1}.$$

Proof In the case of estimation of a location parameter with a known scale parameter, the distribution density becomes a function of one variable denoted as $f(x-\mu)$. From Theorem 2.2 it follows that the MVS score function for μ is of the form:

$$\psi_{MVS}(x-\mu)=\frac{\partial}{\partial\mu}f(x-\mu)+\beta f(x-\mu)=-f'(x-\mu)+\beta f(x-\mu)=-f'+\beta f;$$

here prime is used for differentiation by x, and in the last expression, the arguments are omitted for brevity. The constant β must provide $E\psi_{MVS}=0$:

$$E\psi_{MVS}=-\int_{-\infty}^{\infty}f'f\,dx+\beta\int_{-\infty}^{\infty}f^2\,dx=-\frac{1}{2}\int_{-\infty}^{\infty}d(f^2)+\beta\int_{-\infty}^{\infty}f^2\,dx$$

$$=-\frac{1}{2}f^2\Big|_{-\infty}^{\infty}+\int_{-\infty}^{\infty}f^2\,dx=\beta\int_{-\infty}^{\infty}f^2\,dx=0.$$

Since the last integral is positive, $\beta=0$.

As $\psi_{MVS}=-f'$, we have $\int_{-\infty}^{\infty}\psi_{MVS}^2\,dx=\int_{-\infty}^{\infty}f'^2\,dx.$

Further, $(\partial/\partial\mu)\psi_{MVS}=-\psi'_{MVS}=f''$. Next,

$$E\frac{\partial}{\partial\mu}\psi_{MVS}=\int_{-\infty}^{\infty}f''f\,dx=\int_{-\infty}^{\infty}f\,df'=ff'\Big|_{-\infty}^{\infty}-\int_{-\infty}^{\infty}f'^2\,dx=-\int_{-\infty}^{\infty}\psi_{MVS}^2\,dx.$$

Thus, in the ratio defining VS_{min}, the reduction of terms takes place

$$VS_{min}=\int_{-\infty}^{\infty}\psi_{MVS}^2\,dx\Big/\left(E\frac{\partial}{\partial\mu}\psi_{MVS}\right)^2=\left|E\frac{\partial}{\partial\mu}\psi_{MVS}\right|^{-1}=\left(\int_{-\infty}^{\infty}\psi_{MVS}(x-\mu)^2dx\right)^{-1}.$$

2.4 Definition of Stability

The situation with the measuring of stability is analogous to the situation with the measuring of efficiency. The asymptotic variance V is a positive measure of efficiency, whereas the functional VS is a positive measure of sensitivity. The asymptotic variance attains its minimum V_{ML} at the maximum likelihood (ML) estimator, whereas the functional VS attains the minimum VS_{min} at the minimum variance sensitivity (MVS) estimator. Using VS_{min} as a baseline for the measuring of sensitivity, it is natural to define stability analogously to efficiency in the range from zero to unit.

Definition 2.3 (Shurygin 1994a, b) The *stability* of an M-estimator is defined as the following ratio

$$stb(\psi, f) = \frac{VS_{min}(f)}{VS(\psi, f)}. \tag{12}$$

The optimization of estimation with respect to two indicators on the square *eff*, *stb*—is a way to the solution of the problem of stable estimation. This way depends on the desirable choice of the relation between the values of those indicators: an ML-estimator can be of a low stability, whereas an MVS-estimator can be of a low efficiency. Searching for compromise, the following result may be useful.

Theorem 2.3 (Shurygin 1994a, b) *In the conditions of Theorem 2.2, the maximum of stability under the required efficiency (or the maximum of efficiency under the required stability) are attained at the M-estimator with the following score function*

$$\psi_{c.opt}(x, \theta) = c\left(\frac{\partial}{\partial\theta}\log f(x, \theta) + \beta\right) \Big/ \left(1 + \frac{\gamma}{f(x, \theta)}\right), \tag{13}$$

where the sense of the constants c and β is explained in Theorem 2.2, and the constant $\gamma = \gamma(\theta) > 0$ is defined by the chosen rate of efficiency (or stability).

The proof is based on the application of Lemma 2.1.

Definition 2.4 The M-estimators defined by the score function (13) are called *conditionally optimal.*

Often conditionally optimal M-estimators can be approximated by low-complexity ones. One of these possibilities is given by the following M-estimators.

Definition 2.5 The M-estimator with the score function

$$\psi_{rad}(x, \theta) = c\left(\frac{\partial}{\partial\theta}\log f(x, \theta) + \beta\right)\sqrt{f(x, \theta)} \tag{14}$$

is called *radical.*

As a rule, radical estimators possess the following property: their efficiency is equal to their stability:

$$eff(\psi_{rad}, f) = stb(\psi_{rad}, f). \tag{15}$$

In the case of stable estimation of the location parameter μ (without any loss of generality, we set $\mu = 0$), the *MVS*-, conditionally optimal and radical M-estimators are defined by the following low-complexity score functions:

$$\psi_{MVS} = -f'(x),$$

$$\psi_{c.opt}(x) = \frac{\psi_{ML}(x)}{1 + \gamma/f(x)},$$

where $\psi_{ML} = -f'/f$ is the maximum likelihood score function;

$$\psi_{rad}(x) = \psi_{ML}(x)\,\sqrt{f(x)} = -\frac{f'(x)}{\sqrt{f(x)}}. \tag{16}$$

Note that for the vanishing at infinity densities, all the aforementioned stable M-estimators, namely, *MVS*, conditionally optimal and radical, are redescending.

3 Test Error Sensitivity and Stability

In what follows, we consider the particular case of the tests for location. For the tests of form (2), instead of the asymptotic variance as a measure of accuracy of a test statistic, we use the type II error rate (Shevlyakov et al. 2014)

$$P_E(\psi, f) = 1 - \beta(\psi, f) = \Phi\left(\xi_{1-\alpha} - A\, V^{-1/2}(\psi, f)\right), \tag{17}$$

where $\xi_{1-\alpha} = \Phi^{-1}(1-\alpha)$ (4). Then its Lagrange functional derivative is proportinal to the variance sensitivity of the corresponding test statistic (for details, see (Shevlyakov et al. 2014)).

Definition 3.1 *The error sensitivity* of test (2) is defined as

$$ES(\psi, f) = k\, VS(\psi, f),$$

where k is a constant.

Next, similarly to the tools developed for stable estimation, we search for the optimal *minimum error sensitivity score* minimizing the error sensitivity $ES(\psi, f)$ which evidently coincides with the minimum variance sensitivity score

$$\psi_{MES}(x) = -f'(x) \tag{18}$$

and introduce the test stability as follows.

Definition 3.2 *The stability* of test (2) is set as

$$stb_{test}(\psi, f) = \frac{ES_{min}(f)}{ES(\psi, f)},$$

where $ES_{min}(f) = ES(\psi_{MES}, f)$ is the minimum error sensitivity of a test.

From Definition 3.2 it directly follows that the test stability also coincides with the stability of the corresponding test statistic:

$$stb_{test}(\psi, f) = stb(\psi, f).$$

Thus, due to the simple structure of test (2) and to the natural choice of the type II error rate (17) as a measure of test performance, we can directly apply all the tools developed for stable estimation to the comparative analysis of test performance.

4 Stable Tests for Location

Most results represented below can be found in (Shevlyakov et al. 2014), although some their interpretations have been changed. In what follows, a brief summary of those results is given.

4.1 Problem Setup

Consider testing a location parameter $H_0 : \theta = 0$ versus $H_1 : \theta > 0$ in the model $X_i = \theta + e_i, \quad i = 1, \ldots, n$, where $\{X_i\}_1^n$ are observations, $\theta = A/\sqrt{n}$ is a weak signal, $\{e_i\}_1^n$ are i.i.d. noises with density f.

In this case, the Neyman–Pearson decision rule (2) is used with the M-estimator of location $T_n(X_1, \ldots, X_n)$ as a test statistic satisfying the equation

$$\sum_{i=1}^{n} \psi(X_i - T_n) = 0. \tag{19}$$

4.2 Evaluation Criteria

For performance evaluation of tests, the following criteria, two conventional and one new, are used: (i) the type II error rate, (ii) Pitman's efficiency given by the asymptotic efficiency of a test statistic (an M-estimator $T_n(X_1, \ldots, X_n)$ (19) of a location parameter), (iii) the test stability $stb(\psi, f)$.

4.3 Test Statistic Score Functions

The aforementioned criteria of performance evaluation are used to compare tests with the following score functions: (i) the linear score $\psi_{mean}(x) = x$ with the sample mean as a test statistic, (ii) the sign score $\psi_{med}(x) = \text{sgn}(x)$ with the sample median as a test statistic, (iii) the maximum likelihood score $\psi_{ML}(x) = -f'(x)/f(x)$, (iv) Huber's linear bounded score $\psi_{Huber}(x) = \max[-1.14, \min(x, 1.14)]$ optimal for the model of contaminated Gaussian distributions with the contamination parameter $\varepsilon = 0.1$ (Huber 1964); (v) Hampel's redescending three-part score (Andrews et al. 1972)

$$\psi_{Hampel}(x) = \begin{cases} x, & \text{for} \quad 0 \le |x| \le a, \\ a\ \text{sign}(x), & \text{for} \quad a \le |x| \le b, \\ a\,\dfrac{r-|x|}{r-b}\ \text{sign}(x), & \text{for} \quad b \le |x| \le r, \\ 0, & \text{for} \quad r \le |x| \end{cases}$$

with the parameters $a = 1.31$, $b = 2.039$, and $r = 4$ (Hampel et al. 1986); (vi) the redescending minimum error sensitivity score $\psi_{MES}(x) = -f'(x)$, (vii) the redescending radical score $\psi_{rad}(x) = -f'(x)/\sqrt{f(x)}$.

4.4 Noise Distributions

In the comparative study of tests, the following set of noise distribution densities is used: the standard Gaussian, Laplace, Cauchy, and contaminated Gaussian (Shevlyakov et al. 2014).

The choice of a Gaussian in this list is conventional (Kim and Shevlyakov 2008). The presence of the Cauchy and contaminated Gaussian noise distributions is due to their extremely heavy tails, the detection performance resistance to which is desired. Although the Laplace noise distribution density has moderately heavy exponential tails, it is the least favorable (the least informative) distribution minimizing Fisher information in the class of the nondegenerate distributions (Shevlyakov and Vilchevski 2002): it has the unique combination of a relatively high sharp central peak and relatively heavy tails, and thus it deserves consideration.

4.5 Asymptotic Performance Evaluation

Here we briefly summarize the obtained results (for details, see (Shevlyakov et al. 2014)).

Gaussian Noise: Among robust alternatives, the tests based on redescenders are slightly inferior in efficiency and in the type II error rate to Huber's test. The worst

performance is exhibited by the test with the sign score ψ_{med} asymptotically equivalent to the sign test with the efficiency $2/\pi \approx 0.637$.
Cauchy and Contaminated Gaussian Noises: All the tests based on redescenders outperform Huber's test both in efficiency and the type II error rate. Apparently, the performance of the test with the linear score ψ_{mean} is disastrous in the Cauchy and contaminated Gaussian noises. On the whole, the radical test outperforms the others in the chosen set of noise distributions.
Laplace Noise: In this case, Huber's test performs better than Hampel's one in efficiency and the type II error rate, but the proposed stable tests based on redescenders slightly outperform Huber's detector. Although the Laplace density has lighter tails than the Cauchy, all the competitors are considerably inferior to the maximum likelihood test with the sign score.

4.6 Small Samples Performance Evaluation

In real-life applications mostly small or moderate numbers of observations are available, so the specification of the area of applicability of asymptotic results is of importance. To achieve this, Monte Carlo experiment on samples $n = 20, 40, 60, 80$ and 100 (the number of trials equals 40000) was performed with the decision rule (2) based on the so-called one-step M-estimates. The results of Monte Carlo experiment for the type II error rate can be found in (Shevlyakov et al. 2014). From them it follows that, generally, the small sample results are qualitatively similar to the asymptotic ones.

1. In the Gaussian noise, the best is the ψ_{mean}-test; Huber's and the redescenders except the ψ_{MES}-detector are slightly inferior to it; the worst are the ψ_{med} and ψ_{MES}-tests.
2. In the Cauchy noise, the radical test dominates over the others.

The obtained results confirm that, on the whole, the tests based on redescending score functions outperform Huber's linear bounded conventional tests in heavy-tailed noise distribution models (Shevlyakov et al. 2010).

5 Concluding Remarks

1. Our general aim is to recall some results within Shurygin's approach to stable estimation and to apply them to robust hypothesis testing in the particular case of the Neyman-Pearson decision rules: it is shown that this approach works quite well in heavy-tailed distribution models, outperforming Huber's and Hampel's optimal tests. The aforementioned advantages of Shurygin's approach most reveal themselves in robust estimation and hypothesis testing for location problems. Our

recent experience to apply this approach to stable estimation of a scale parameter has shown that although the designed stable procedures perform rather well, there exist considerably better in efficiency (about 10 %) highly robust estimators of scale, say, the one based on the proposed by Rousseeuw and Croux (1993) highly robust and efficient Q_n-estimator of scale. However, the application of the variational optimization approach to the design of stable estimators of regression, correlation, covariance etc. (especially to the problems of multivariate statistical analysis) deserve a thorough study.

2. The following connection of the variance sensitivity *VS* with the change-of-variance function *CVF* is of importance. For M-estimators of location with continuously differentiable score functions $\psi(x)$, the change-of-variance function is given by (Rousseeuw 1981; Hampel et al. 1986)

$$CVF(x;\, \psi, f) = \frac{A(\psi, f)}{B^2(\psi, f)} \left(1 + \frac{\psi^2(x)}{A(\psi, f)} - 2\frac{\psi'(x)}{B(\psi, f)}\right),$$

where

$$A(\psi, f) = \int \psi^2(x) f(x)\, dx\,, \quad B(\psi, f) = \int \psi'(x) f(x)\, dx\,.$$

Thus, up to an additive constant, the variance sensitivity is equal to the integral of the *CVF*, if it exists (Shevlyakov et al. 2008).

3. Now we consider another way to justify the use of the Lagrange derivative (10) in the case of location

$$VS(\psi, f) = \frac{\partial V(\psi, f)}{\partial f} = \frac{\int \psi(x)^2\, dx}{\left(\int \psi'(x) f(x)\, dx\right)^2}.$$

Assume continuous differentiability of score functions $\psi(x)$ and require that $f(x) \to 0$ as $|x| \to \infty$. Now we analyze the relation of the Lagrange derivative to the variation $\delta V(\psi, f)$ of the asymptotic variance $V(\psi, f)$ finding its principal part with respect to $||\delta f||$ (assume additionally that the admissible variations of densities satisfy $\int \delta f(x)\, dx = 0$):

$$V(\psi, f + \delta f) = \frac{\int \psi^2 (f + \delta f)\, dx}{\left(\int \psi' (f + \delta f)\, dx\right)^2}$$

$$= \frac{\int \psi^2 f\, dx + \int \psi^2 \delta f\, dx}{\left(\int \psi' f\, dx\right)^2 + 2 \int \psi' \delta f\, dx \int \psi' \delta f\, dx + \left(\int \psi' \delta f\, dx\right)^2}$$

$$= \frac{\int \psi^2 f\, dx + \int \psi^2 \delta f\, dx}{\left(\int \psi' f\, dx\right)^2} \left(1 - 2\frac{\int \psi' \delta f\, dx}{\int \psi' f\, dx}\right) + o(||\delta f||)$$

$$= V(\psi,f) + \frac{\int \psi' f\,dx \int \psi^2 \delta f\,dx - 2\int \psi^2 f\,dx \int \psi' \delta f\,dx}{\left(\int \psi' f\,dx\right)^3} + o(||\delta f||).$$

Finally, we get

$$\delta V(\psi,f) = V(\psi,f+\delta f) - V(\psi,f)$$

$$= \frac{B(\psi,f)\int \psi^2 \delta f\,dx - 2A(\psi,f)\int \psi' \delta f\,dx}{B(\psi,f)^3} + o(||\delta f||)$$

$$= \int \left(\frac{\psi(x)^2}{B(\psi,f)^2} - 2\frac{A(\psi,f)\psi'(x)}{B(\psi,f)^3} \right) \delta f(x)\,dx + o(||\delta f||).$$

Taking into account the analog of the Lagrange mean theorem (Bohner and Guseinov 2003), we arrive at the following relation for the sought variation

$$\delta V(\psi,f) = \delta f(x^*) \int \left(\frac{\psi(x)^2}{B(\psi,f)^2} - 2\frac{A(\psi,f)\psi'(x)}{B(\psi,f)^3} \right) dx$$

where $x^* \in \mathbb{R}$, and thus its principal part is proportional to the Lagrange derivative

$$\delta V(\psi,f) \propto \frac{\partial V(\psi,f)}{\partial f}.$$

In other words, the Lagrange derivative reflects the variation of the asymptotic variance $V(\psi,f)$ corresponding to the uncontrolled variation of density f.

4. Now we focus on the statistical sense of the variational optimization approach to stable estimation. Originally, it does not involve the consideration of neighborhoods of an ideal distribution model, like in the historically the first Huber's minimax approach to robust estimation at ε-contaminated distributions (Huber 1964, 1981). In Huber's approach, the contamination parameter ε is in essence unknown, otherwise we could use the maximum likelihood estimator in the ε-contaminated model. However, even within the variational optimization approach, it is possible to introduce a neighborhood of density f and to design minimax variance sensitivity estimators—such conventional redescenders like Smith's and Tukey's biweight have been derived in (Shevlyakov et al. 2008).
Another situation is with Hampel's approach based on influence functions: the variational optimization approach is close to that (in no way we compare the levels of the elaboration of these concepts) yielding not local indicators of robustness as the influence and change-of-variance functions, but a global indicator of stability of estimation in some aspects similar to the notion of efficiency of estimation.

Acknowledgments I would like to thank the referees for their important and helpful comments. Also I am so much grateful to Prof. Ayanendranath Basu for his patience and help, which allowed me to finalize this paper.

References

Andrews DF, Bickel PJ, Hampel FR, Huber PJ, Rogers WH, Tukey JW (1972) Robust estimates of location. Princeton Univ. Press, Princeton

Bentkus V, Bloznelis M, Götze F (1995) A Berry-Esseen bound for M-estimators. Preprint 95-068, Universität Bielifeld

Bohner M Guseinov GSH (2003) Improper integrals on time scales. Dyn Syst Appl 12:45–65

Hampel FR (1974) The influence curve and its role in robust estimation. J Am Stat Assoc 69:383–393

Hampel FR, Ronchetti E, Rousseeuw PJ, Stahel WA (1986) Robust statistics. The approach based on influence functions. Wiley, New York

Hossjer O, Mettiji M (1993) Robust multiple classification of known signals in additive noise - an asymptotic weak signal approach. IEEE Trans Inf Theory 39:594–608

Huber PJ (1964) Robust estimation of a location parameter. Ann Math Statist 35:1–72

Huber PJ (1981) Robust Stat. Wiley, New York

Kim K, Shevlyakov GL (2008) Why Gaussianity? IEEE Signal Process Mag 25:102–113

Michel R, Pfanzagle J (1971) The accuracy of the normal approximation for minimum contrast estimates. Z. Wahrsch Verw Geb 18:73–84

Noether GE (1967) Elements of nonparametric statistics. Wiley, New York

Rousseeuw PJ (1981) A new infinitesemal approach to robust estimation. Z Wahrsch Verw Geb 56:127–132

Rousseeuw PJ, Croux C (1993) Alternatives to the median absolute deviation. J Amer Statist Assoc 88:1273–1283

Shevlyakov GL, Vilchevski NO (2002) Robustness in data analysis: criteria and methods. VSP, Utrecht

Shevlyakov GL, Morgenthaler S, Shurygin A (2008) Redescending M-estimators. J Stat Plann Infer 138:2906–2917

Shevlyakov GL, Lee JW, Lee KM, Shin VI, Kim K (2010) Robust detection of a weak signal with redescending M-estimators: a comparative study. Int J Adapt Control Signal Proc 24:33–40

Shevlyakov GL, Shin VI, Lee S, Kim K (2014) Asymptotically stable detection of a weak signal. Int J Adapt Control Signal Proc 28:848–858

Shurygin AM (1994a) New approach to optimization of stable estimation. In: Proceedings 1st US/Japan Conference on Frontiers of Statist. Modeling, Kluwer, Netherlands, pp 315–340

Shurygin AM (1994b) Variational optimization of the estimator stability. Autom Remote Control 55:1611–1622

Shurygin AM (2009) Mathematical methods of forecasting. Text-book, Hot-line-Telecom, Moscow (in Russian)

Tukey JW (1977) Exploratory data analysis. Addison-Wesley, Reading

List of Referees

The following individuals have acted as referees for the articles submitted for publication in this volume. The editors express their deep sense of gratitude to all of them.

Arslan, Olcay
Abel, Guy
Bellio, Ruggero
Bhar, Lalmohan
Bortot, Paola
Carioli, Alessandra
Chenouri, Shoja'eddin
Croux, Christophe
Das, Rabindra Nath
Di Marzio, Marco
Genton, Marc
Ghosh, Abhik
Greco, Luca
Hennig, Christian
Jain, Kanchan
Jamalizadeh, Ahad
Kalina, Jan
Kent, John
Khalili, Abbas
Kuchibhotla, Arun Kumar
Leung, Andy
Ley, Christophe
Mandal, Abhijit
Marazzi, Alfio
Markatou, Marianthi
Masarotto, Guido
Mayo Iscar, Agustin
Monti, Anna Clara
Morgan, John P.
Morgenthaler, Stephan
Nordhausen, Klaus
Oja, Hannu
Öllerer, Viktoria
Pardo, Leandro
Riani, Marco
Raymer, James
Rousseeuw, Peter
Ruckdeschel, Peter
Shevlyakov, Georgy
Singer, Julio M.
Vogel, Daniel
Zamar, Ruben

C. Agostinelli et al. (eds.), *Recent Advances in Robust Statistics: Theory and Applications*, DOI 10.1007/978-81-322-3643-6

MIX
Papier aus verantwortungsvollen Quellen
Paper from responsible sources
FSC® C105338

If you have any concerns about our products,
you can contact us on
ProductSafety@springernature.com

In case Publisher is established outside the EU,
the EU authorized representative is:
Springer Nature Customer Service Center GmbH
Europaplatz 3, 69115 Heidelberg, Germany

Printed by Libri Plureos GmbH
in Hamburg, Germany